THE TRANSPORTATION OF OIL BY SEA

Also by the same author
and available from iUniverse, Inc.
Selected Issues in Agricultural Policy Analysis
with Special Reference to East Africa

THE TRANSPORTATION OF OIL BY SEA

Tony Akaki

iUniverse, Inc.
Bloomington

THE TRANSPORTATION OF OIL BY SEA

iUniverse books may be ordered through booksellers or by contacting:

iUniverse
1663 Liberty Drive
Bloomington IN 47403
www.iuniverse.com
1-800-Authors (1-800-288-4677)

ISBN-13: 978-0-595-36545-6 (pbk)
ISBN-13: 978-0-595-80976-9 (ebk)
ISBN-10: 0-595-36545-0 (pbk)
ISBN-10: 0-595-80976-6 (ebk)

Printed in the United States of America

DEDICATED TO DORA

Contents

1

INTRODUCTION

Bulk shipping has been used for many years to reduce the cost of sea transport; two thousand years ago Rome imported more than thirty million bushels of grain a year from the grain baskets of northern Africa, Sicily, and Egypt. To carry this trade, a fleet of special ships was built. However, it was not until the mid-nineteenth century that the oil tanker became an important element in the trading of oil.

The beginnings of the tanker trade may be found in the history of oil, and particularly in Titusville, Pennsylvania, where the world's first major oil field was discovered. The place became a boomtown; with the leases on promising land changing hands for thousands of dollars, fortunes were being made overnight. The demand for the new product was insatiable, and by the end of 1861, kerosene made from oil had virtually displaced coal as the main source of fuel. Oil also began to make an important contribution to the war effort of the Union in the United States' civil war; the North needed the new source of foreign exchange to compensate for the loss of the South's cotton.

Oil exports from America began under interesting circumstances in December 1861 when the *Elizabeth Watts* sailed from Philadelphia to London loaded with barrels of kerosene. The transportation of oil has its inherent risks and thus has always been a fact to be addressed, and in this regard it is important to note that the crew of the *Elizabeth Watts* were apprehensive in loading the cargo of kerosene for **fear** of fire. Drunkards from the bars along the Delaware River were found to load the vessel that marked the birth of the transportation of liquid fuel on a major scale.

The voyage was a success, and by the end of 1865 Britain, France, and Germany were all substantial buyers, and most of the nineteenth-century exports accounted for over a third of U.S. annual production. Nevertheless, it became clear that the transportation of oil barrels in wooden vessels was not only dangerous but also uneconomical. This led to the building of a vessel in which the hull or skin of the ship acted as a container; the vessel was called the *Zoroaster* and was followed very shortly by the steamer *Glueck-Auf.*

The *Glueck-Auf,* which was built in 1886 in the United Kingdom, is now generally accepted as the prototype of the modern oil tanker. It is significant to note here that there are very few similarities between it and some of the modern tankers. First, transportation of oil in the early days consisted largely of products such as kerosene. There was, of course, hardly any use for petroleum at the time, and the transportation of crude oil did not start until some considerable time later.

Unlike gold, oil does not have intrinsic value, and when production outruns demand prices are bound to fall; early speculators overlooked this vital fact when they paid vast sums for tracts of promising land. During the 1860s prices dropped from the original twenty dollars a barrel to two dollars, and during the United States' civil war they fluctuated wildly from ten cents a barrel, which made the wooden barrels more valuable than the oil they contained, up to fourteen dollars a barrel. After the Civil War the day of the small oilman was over and the era of big business started. On the January 10, 1870, the Standard Oil Company was incorporated in Cleveland with John D. Rockefeller as its president. Rockefeller was determined to become the largest shipper of oil.

But before he reached a position to exert any real leverage over the industry, the leading railroads took the initiative out of his hands. However, the economic depression, also known as the Long Depression, strongly affected the transport industry in 1873, making it easy for Rockefeller to approach other competitors and players in the industry and take them over. He eventually succeeded in forming a monopoly in the transportation of oil.

Other parts of the world were also developing viable interests. Baku, in the Caucasus, had been seeping oil through the Earth's surface for hundreds of years. The Nobel brothers quickly established an ascendancy over Baku and, following Rockefeller's example, secured a virtual monopoly over transportation of oil from the Caucasus to the rest of Russia and abroad.

This was a complicated process, as there was no railway to the Black Sea, which was the logical output for exports from Baku. The oil had to be taken up the Volga and thence by rail to the Baltic. Other operators tried to build a railway to the Black Sea port of Batum, but the Nobels promptly slashed the price of their oil to levels so low that it became impossible for other competitors to raise the necessary finance. The only alternative was to raise the necessary funds from outside of Russia if the project was to be undertaken. Accordingly, in 1880 the Russians approached Baron Alphonse de Rothschild in Paris, France.

Rothschild's Paris bank was already involved in oil and consequently put up the money for the Batum railway, but in return they demanded mortgages on the producer's properties as well as the right to buy oil for export. Nevertheless, with the completion of the railway in 1883 the growth of the Russian oil industry was incredible, and by 1888 its production was over two and a half million tons.

This intermodal characteristic of the oil industry between sea and rail transport was also significant in 1860s America, particularly in Cleveland, where Rockefeller had set up the Standard Oil Company. Cleveland had a direct rail link with the oil regions in Pennsylvania, combined with unrivaled communications with the eastern United States by two railways and the Great Lakes with their interlocking system of canals.

This intermodal nature had a huge impact on oil distribution. For instance, Russian oil swept all over Europe like a flood and transformed the entire marketing situation. By the 1890s Russian oil could be shipped via the Suez Canal to the Far East more easily than from North America and afterwards the Dutch East Indies became a major producer, and Burma a less important one. It is important to note the considerable impact of the Suez Canal on oil transport once the movement of crude oil had reached relatively large quantities.

Interestingly, the first vessel to transit the Suez Canal fully laden belonged to one of the oil majors': the Royal Dutch Shell Group's the *Murex*. On August 24, 1892, it was the first of several tankers to sail from the shipbuilder's yard in West Hartlepool, United Kingdom, for the Black Sea, and a few weeks later it discharged its cargo in Singapore and Bangkok. Long and arduous negotiations with canal authorities were necessary before the ship could undertake this first voyage from the Black Sea to the Far East.

This voyage through the canal from north to south remained the norm until the beginning of the twentieth century. Contrary to common belief that the situation changed with the discovery of oil in the Arabian Gulf (which is also known as the Persian Gulf), it was the exploration and eventual production of Indonesian oilfields that opened up a trade for oil from the East through to Europe. Until that time, ships carried oil southward through the Suez and carried dry cargo from the East on the return journey. The possibility of carrying oil in both directions led to innovative tanker designs. It is also crucial to point out that, since 1975, the Suez Canal Authority has been aware that fully laden Very Large Crude Carriers (VLCC) are unable to transit the canal; however, discussions on deepening the canal sufficiently for VLCCs to transit while fully laden are ongoing.

The competition in the Far East was intense, for the Far Eastern markets were dominated by Standard Oil, and any newcomer trying to break into the area would become an easy victim to the kind of selective price cutting by which Standard Oil maintained its supremacy in the United States The practices of Standard Oil and other companies involved in oil has brought a lot of attention to the business of oil production, marketing, and transportation, as well as what may be called the politics of oil.

However, this book is essentially about the transportation of crude oil and not the politics of oil. Oil is a highly emotive issue that involves governments, powerful finance houses, and oil companies, among others. That notwithstanding, it is therefore inevitable that some political, geostrategic, and economic aspects of oil transportation will be examined, particularly as they reflect on the vulnerability of Western interests in the Middle East.

Royal Dutch Shell Group

Competition among the major oil companies that had become multinationals and wielded enormous power signaled the beginning of the tanker chartering era. In April 1907 two of the largest European companies—Royal Dutch Petroleum Company, under Henri Deterding, and Marcus Samuel's "Shell" Transport and Trading Company—merged into the Royal Dutch Shell Group.

The fortunes of Shell are interesting to those involved in the transportation of oil by sea.

The tanker fleet of Shell alone was more numerous than the merchant marine of all but the largest countries. This era of competition, characterized by cutthroat price wars, is what led to the amalgamation of these giants; smaller competitors were pushed out of the market, thereby consolidating power into the hands of a few large multinationals. The Shell company is of particular interest to those involved in the tanker trade, not only because of its large fleet of tankers but also for the innovations it brought to the tanker trade in general.

Shell's business was conducted through a network of agents in the main trading centers, and instead of owning their own vessels they chartered vessels through a firm of London brokers called Lane & MacAndrew. One of its partners was Fred Lane, who was also the London agent for the Paris Rothschild's oil interest. He wanted to find a way of selling their Russian kerosene in the Far East, and he suggested carrying out this project to Marcus Samuel sometime between 1885 and 1888.

Fred Lane knew that only a company capable of establishing itself simultaneously in every market could hope to survive. Lane argued that this was possible with their existing network. Marcus Samuel was skeptical because of oil's inflammability; furthermore, kerosene could only be taken through the Suez Canal when packed in tins that were in turn packed in wooden cases. Both the tinplate, which came from Wales, and the wood were very expensive, Samuel wondered whether it would be possible to match Standard's prices. Nevertheless, he was sufficiently interested to allow Lane to take him on a tour of the Russian oil industry by 1890. At Batum, the Black Sea port from which Russian oil was exported as mentioned earlier, Samuel came up with an idea that would transform the transportation of oil.

Although the Russians were the first people to develop seagoing tankers able to carry oil in bulk, safety issues were a huge consideration on those early voyages, and vessels blowing up were common events. It was here that Samuel came up with the aforementioned idea that revolutionized the transportation of oil. He knew that if a tanker could be designed to satisfy the safety requirements of the Suez Canal Authority, he could take kerosene through the canal in bulk, and then, after being steam-cleaned, return to Europe carrying rice, tea, or some other cargo. Under those circumstances, his agents could easily undercut the Standard Oil Company.

This state of affairs is what led to the building of the *Murex* and her sister ships, but only after the Suez Canal Authority gave their blessing to the plans. With this guarantee the Rothschilds agreed to finance the project. The Samuel's cost calculations were proved correct, and the oil was put on sale at less than half the price of Standard Oil's. Nevertheless, despite its cheapness nobody seemed to want it. In Europe buyers brought their own containers in which to carry their kerosene, but in Asia the battered old tins in which Standard Oil's product traveled were also used to build homes or were refashioned into cups, plates, buckets, and hundreds of other things. Therefore, several more months had to be wasted while container factories were built near the storage depot to facilitate trade in Asia.

Furthermore, the tankers were the only part of the enterprise Samuel actually owned himself. For the rest of the business he operated a syndicate, known as the tank syndicate, in conjunction with his agents in the Far Eastern ports; all transactions concerning both the kerosene for the outward journeys and cargoes for the return trip were conducted on behalf of a joint account of all members. In this way Standard Oil was prevented from picking them off one by one through selective price-cutting or takeovers.

By 1897 the Shell Transport and Trading Company was formed to take over the entire operation, with members of the syndicate becoming the first shareholders. The formation of this company was in response to the booming business of oil production, marketing, and transportation.

However, this period was the lull before the storm. Since then, tanker shipping has proven to be the most vulnerable type of shipping during an international crisis. World freight markets have proved to be very sensitive to world events and have had huge impacts on tanker operators. This is best illustrated by highlighting the events that affected the Shell Transport and Trading Company.

For instance, The Boxer Rebellion and the siege of Peking (1898–1900) ruined the Chinese market, where Shell had built up large stocks. Meanwhile in Russia, the onset of an economic slump forced the refineries there to switch sales to the export market. Consequently, as supply increased, prices everywhere collapsed.

The situation would have seemed hopeless had it not been for the discovery of the great Spindletop field in Texas on of January 10, 1901, which at the time was the biggest gusher recorded. On hearing the news, Shell opened negotiations with Gulf, the oil producers and owners of Spindletop, and prepared for another coup that could potentially make Shell as powerful as it was in the Far East. A massive contract involving a minimum delivery of ten thousand tons a year at a fixed price for the next twenty-one years was signed with Colonel Guffey, the founder of Gulf. Four new ten thousand ton tankers were designed and built to bring the oil across the Atlantic. In England, Shell hoped to persuade a larger importer to join the operation and eventually formed a partnership with Deutsche Bank.

Soon the Shell tankers were bringing cheap oil from Texas to Europe. Samuel knew little about oil exploration, and he never had a dominant position in European markets to which Standard could ship oil cheaply across the Atlantic. However, Samuel was a shrewd trader and shipper of oil, and this enabled him to establish a company in a Standard Oil stronghold: Germany.

In August 1902 Spindletop stopped gushing after only twenty months. Within a few months Shell's supplies from Texas were completely cut off, leaving the company with large contracts to fulfill but no means to do so. There was no longer any work for the ten thousand ton tankers. The pride of its fleet had to be converted into cattle carriers.

Shell had no recourse but to continue to buy expensive Russian and Far Eastern supplies to meet its European commitments, and three years later, in 1905, the situation became untenable. It was forced into a humiliating withdrawal from the continental markets when it had to sell off its best tankers to the Deutsche Bank as the price of getting out of their partnership agreement.

The fortunes of the Shell Transport and Trading Company highlighted heretofore clearly portray how unpredictable the oil business can be. It is also fundamental to note the influence of a huge monopolistic power—Standard Oil—that was not afraid to throw its weight around in its desire to further expand and dominate the oil business. While Shell's tankers lay idle, Standard Oil continued to cut prices and expand into Europe, putting up a factory in Romania and eventually in Germany, where Shell was ousted through the machinations of its erstwhile partner Deutsche Bank.

Standard Oil did not succeed in taking over Shell, although it did offer to buy out Shell for forty million dollars, promising a joint subsidiary with Samuel as chairman. Samuel, however, preferred to merge with Royal Dutch, where he thought he could be master, but after Spindletop dried up he was forced to discuss a total merger with Deterding, on humiliating terms of sixty-forty and with Deterding as the manager director.

As already mentioned, the tanker trade is one of the most vulnerable types of shipping. It also goes without saying that, precisely because it is a risky business, some shrewd businessmen have been able to enter this volatile business and make a success of it. Before the coming of the dot-com era, it was by no accident that the richest individuals in the world were involved with tankers. Probably the most famous ship owner in the world was Aristotle Socrates Onassis. Before one looks at the tanker freights from which Onassis, among others, managed to secure a good return on capital, it is crucial to recap and examine some of the factors that affected the tanker industry, starting with World War II. The war spurred the building of the standard vessel types such as the T2.

From 1942 to 1945, nearly five hundred T2s were built. At the beginning of the war, the world tanker fleet was around sixteen million tons deadweight. At the end of the war, in spite of war losses, the world tanker fleet had risen to around twenty-four million tons deadweight. The period from 1966 to 1973 marked the most rapid expansion of both the oil industry and the tanker trade.

This period brought about the first generation VLCCs (i.e., vessels up to 300,000 tons deadweight—anything over 300,000 tons is known as an Ultra Large Crude Carrier, or ULCC). The total tonnage that existed in 1966 was 97 million tons deadweight. This increased over the next seven years to over 220 million deadweight tons. These are figures are for pure tankers only. If the tonnage from combination, or OBO (Ore/Bulk/Oil) carriers were included, another 37 million deadweight tons ought to be added to the tanker fleet.

The year 1973 was significant to the tanker operation business as a result of the Yom Kippur War in October. Until that time prices had remained below two dollars a barrel after prices rose briefly after the initial discovery of oil. Prices rose to thirty-four dollars a barrel in the early 1980s before briefly declining to around ten dollars before rising again during the Gulf War in the early 1990s.

Before the events of 1973 that led to increased in oil prices, tanker operators in expectation of windfall profits from an insatiably oil hungry world embarked on the biggest shipbuilding spree in history. A reduced demand for oil was caused not only by price but also by an economic recession that reduced the demand for tankers. The result was a surplus of hundreds of ships. This is in light of the fact that 1972 was a depressed year for all consumers in the developed world. The world's oil consumption was 2.6 billion tons of which 55 percent was moved by oil tankers; aside from Japan's ninety percent, this included 60 percent of Europe's oil and 60 percent of Australasia's. It comes as no surprise to see how devastating the war was on the world economy.

Many VLCC orders were cancelled as a result of the decline in oil demand that followed the fourfold increase in oil prices in 1974 and widespread economic recession in the industrial world. Nevertheless, the world tanker fleet remains dominated by hundreds of massive ships and will remain so for the foreseeable future. Giant oil tankers will continue to move most of the world's oil, and, most importantly, they will handle much of the United States' imported oil.

2

TANKER FREIGHTS

The majority of tanker contracts are arranged in accordance with the terms and conditions of the specialized Worldwide Tanker Nominal Freight Scale, or Worldscale for short. This scale applies to tankers carrying oil in bulk. Although intended solely as a means of reference by which all voyages and market levels could be compared and easily judged, it has developed into the basis of freight assessment for the tanker operation business.

Some people outside the tanker community have questioned the need for Worldscale and cannot see why people in the tanker industry do not agree on a dollar-per-ton rate in exactly the same way as for dry cargo. Of course, there is no reason why two parties to a tanker voyage charter should not agree on a dollar-per-ton freight rate. However, given the nature of oil, this is likely to lead to severe complications. Oil is a relatively homogenous commodity, and the buying and selling of crude oil in particular rarely allows the cargo owner to arrange a sale in advance. The cargo owner is unlikely to know the final destination of the cargo at the time the vessel is fixed, even after the ship has sailed from loading port.

In the dry cargo trades one at least knows the country of destination before a charter is fixed. In the oil trade, charters of vessels need as much flexibility as possible with regard to discharging ports. It is virtually impossible in any charter party contract to agree to a specific rate for every port of discharge that the charterer might wish to use, and one can only imagine how long it would take to fix a vessel if one had to negotiate a separate dollar to freight rate, for example at ten different discharging ports.

This is the main reason why Tanker Freight Scales, which established certain basic rates in advance, came into existence. There have been a number of Tanker Freight Scales in existence since World War II, and in the United States a particular form of scale is used for its national flag carriers. This scale is known as ATRS (American Tanker Rates Scale).

The general purpose of all Tanker Freight Scales has been the same since their introduction during World War II. In January 1989 New Worldscale (now simply Worldscale) was introduced. This new version made several alterations to previous scales and will be the basis of further discussion on the matter of tanker freight rates

The Structure of Worldscale

Worldscale is structured to produce a rate per ton for every possible combination of loading and discharging ports throughout the world that will give an equivalent return, regardless of the voyage itself. There are about sixty thousand actual rates in Worldscale, and when these are varied with the addition system, over two million individual voyages can be rated. It is obviously impossible to fix a scale for every vessel that exists. Therefore, it becomes imperative to have what is known as a *standard vessel.*

The standard vessel is used for the purpose of calculating the basic rates only. The standard vessel of choice of Worldscale is a 75,000 tonner, which is much more representative of the average size crude tanker. Therefore, a more balanced basis of rates is possible, while the standard vessel of the previous Worldscale was a 19,500 tonner. The differences in size do not invalidate the efficiency of either scale because in each case the standard vessel was used in calculating all the rates within the scale to which it applied.

Worldscale is updated on the first day of each calendar year, and this reflects changes associated with port charges and bunker prices. Regarding bunker prices for fuel oil, Worldscale normally takes the prevailing price during the previous September before the scale comes into operation; port charges are also calculated during the previous September, and the Worldscale Association shows the exchange rates used in the various national ports against the U.S. dollar. Furthermore, the round voyage distance is also given against every voyage in the scale; therefore, in order to find the distance between the loading and discharging port, it is necessary to halve the distance given.

Now that we have the description of the standard vessel, the distance, the cost of bunkers, and the cost of port charges—combined with any canal dues that might be transited during the voyage—it is now possible to work out the earnings of a ship. Worldscale assumes and wishes every voyage rate to show a return of twelve thousand U.S. dollars per day. This is called the *fixed hire* element in the basis of calculation. What this means is that the rate given in Worldscale should earn twelve thousand dollars for the vessel for every round voyage in the scale.

However, the Worldscale Association stresses that twelve thousand dollars does not represent a reasonable return on capital that covers operating costs; it is merely a dollar figure used to work out the various basic rates. In regard to canal transits, Worldscale allows twenty-four hours for every transit of the Panama Canal and thirty hours for every transit of the Suez Canal. Canal transits cost money to the ship owner. This expense is built into the freight rate, and this is achieved by awarding the ship owner a fixed rate differential for every transit of such a canal.

Worldscale covers only two canals—namely, the Panama and Suez Canals (the St. Lawrence Seaway/Welland Canal is a special case)—and are dealt with in a slightly different way. For the Panama Canal there are two rates, which are calculated on the vessel's Panama Canal Net Tonnage (PCNT); one rate applies when a vessel transits the Panama Canal in ballast only, and the second and higher rate is payable if the transit involves both a ballast and laden transit of the canal.

The Suez Canal rates are a little more complicated, but are also based on two levels of rates depending on whether the fixture involves ballast only or a ballast and laden transit of the canal. However, because of the considerable spread of tonnage that transits the canal, it was deemed necessary to divide the vessels into four categories that represent the vessel's Suez Canal Net Tonnage (SCNT). These are:

5,000–10,000 tons
10,000–20,000 tons
20,000–85,000 tons
Over 85,000 tons

For the first two categories, the rate is identical for crude or products, but for the categories over 20,000 tons there are different rates:

a. For crude only

b. Products, or crude and products

Transit through the Suez Canal also involves both a dollar-per-ton payment and an additional lump sum, while the Panama Canal transit is covered by a single dollar-per-ton payment. With all this information readily available in Worldscale it is now possible to check out the basic rate for every voyage provided.

Furthermore, Worldscale conditions specify that port charges (e.g., quay dues) at certain ports are taken to be for the account of one or the other of the contracting parties, and agreement to abide by Worldscale terms and conditions implies agreement to accept the laid-down division of these expenses. Worldscale makes no such allowance for freight taxes, however, and as for dry-cargo fixtures, the parties to a tanker negotiation must agree which of the two will ultimately be responsible for such changes.

As already mentioned, the result of the voyage estimate containing the above factors is expressed in terms of U.S. dollars per ton, and described as being equivalent to a scale rate of Worldscale 100 (W/S 100). Market forces dictate whether the level for any particular voyage for any size vessel should be higher or lower than W/S 100, and fixtures may be made at, say, W/S 50 or W/S 200 or whatever depending upon supply and demand at the time of fixing the vessel.

At W/S 50, half the published freight rate to the fixture, while at W/S 200, twice the published freight rate is applicable. For example, if a voyage from Curacao to New York were shown in the Worldscale Freight Scale Book as $5.44 per long ton, a fixture at W/S 50 would equate to a freight rate of $2.72 per long ton against the cargo; likewise, for a fixture at W/S 200, this freight rate would be $10.88.

Average (Tanker) Freight Rate Assessment (AFRA)

In concluding the matter of freights it is important to state that, in much the same way as shipbrokers are involved in the organization of Worldscale, there is also a separate body of shipbrokers that forms the committee of AFRA. This is an independent panel, although it was originally set up for the benefit of oil companies.

During the 1960s and 1970s, when the major oil companies dominated the market more than they do today, there were frequent dealings between them for freight. Furthermore, these oil majors were also big ship-owning entities that had dealings between themselves in re-letting their vessels to each other, usually off the market. In order to avoid the hassle of arguing about the appropriate rate of freight on every occasion, they decided to set up the AFRA organization, whose rates they could then use whenever they required an interchange of tonnage.

The AFRA panel meets to assess the current market levels and to publish its own view of the average of such rates. These rates are one of the fairest ways of deciding the market level at any given time. The oil companies then accepted these rates as governing the various charter parties that were concluded between them, and this saved a great deal of unnecessary negotiations. However, since the oil majors are disinvesting in the ship-owning business such rates are not frequently used.

Tanker and Oil Market Overview

Tanker owners enjoyed the rewards of an oil production boom in 1988 as OPEC members fought a political war over the price of crude and production ceilings. The end of the Iran-Iraq war after nearly eight years was the biggest event for both the tanker and oil markets. This is reflected from a 1987 base output of 1.7 million barrels of oil per day, to production levels of 23 million barrels by October 1989.

VLCCs were able to command October and November 1989 payments of over Worldscale 100 for voyages west and over Worldscale 90 going east. It was welcome relief as the mid-year position was stagnant, with payments between Worldscale 40 and 50.

Rate of scrapings reduced to a trickle in 1989 as owners held out for the long awaited upturn, encouraged by the strength of the secondhand market. In the first three quarters of 1989 a mere 1.1 million tons of wet trading tonnage found its way to the scrap yards; by comparison, 1988—acknowledged as an appalling year for scrapping—saw 2.5 tonnes scrapped, a far cry from figures of 7.8 million tons, 14.5 million tons, and 29.4 million tones recorded in the three preceding years.

Yet the average life span of a VLCC is 12.7 years, and 725 of the biggest vessels were built before 1978. Theoretically, these ships should be scrapped from the mid-1990s onwards, but many owners, driven by the high cost of replacement, are seeking to lengthen the life of their tankers.

Leading the way was Shell, with the revelation in May 1989 that it was to carry out extensive refits to its 4 L-class ULCCs built in the late 1970s to extend their service life for another ten years. In spite of a steep rise in new building prices, the surge in tanker rates at the end of 1988 encouraged a flood of new building orders, especially in the first half of the year. A researcher at shipbroker John I. Jacobs reported that the order book for tankers at the end of June was the highest in nine years.

Orders for 59 tankers totaling 6.1 million tonnes were placed during the first six months of 1989, almost twice the volume of contracts seen in the half of 1988. The world order book stood at 215 tankers of 19.2 million tones, up 2 million tonnes on the beginning of the year.

3

SHIP REGISTRATION/FLAGS OF CONVENIENCE

By Onassis' own account, he was responsible for one of the major innovations of the post-Depression years; namely, the systematic exploitation of flags of convenience has now become a permanent feature of shipping in general. Although Onassis did not start flags of convenience, he made it popular.

Before proceeding further, it is useful to clarify a couple of definitions, such as *national flag* and *flag of convenience*. In shipping circles, a national flag carrier denotes a vessel belonging to an established maritime nation and is mainly manned by seamen of that nation who are almost certainly members of a national trade union.

It is also important to note that flying a national flag does not necessitate that the ownership of the vessel be resident in the country. National flag countries are characterized by established merchant fleets. Examples include: the United States, Japan, Greece, Great Britain, and Denmark. On the other hand, a flag of convenience signifies a country without a significant maritime history. Countries such as Panama and Liberia are flags of convenience.

The movement toward international open registers of flags of convenience started in the 1920s, when U.S. ship owners saw registration under the Panamanian flag as a means of avoiding the high tax rates in the United States while at the same time registering in a country within the stable political orbit of the United States.

The switch to flags of convenience can be illustrated by the fact that 89 tankers were registered in Panama in 1947, and by 1948 the number had risen to a 173. In 1970, 20 percent of the world tanker fleet flew flags of convenience, and a decade later this rose to 35 percent. This was due to newly emergent countries also building fleets. The two tables that follow illustrate this trend.

Table 1 **Changes in World Merchant Fleets (Number of Ships)**

Country	1956	1966	1976	1980
France	1,220	1,558	1,393	1,247
Greece	350	1,377	2,743	3,827
India	221	354	471	601
South Korea	n/k	150	828	1,287
Japan	1,770	5,836	9,932	9,981
Liberia	582	1,287	2,520	2,466
Norway	2,351	2,742	2,706	2,531
Panama	555	692	2,418	3,803
Singapore	-	22	610	1,031
Turkey	278	287	387	475
UK	5,632	4,437	3,622	3,211
United States	4,102	3,111	4,346	5,088

Source: Lloyd's Register

Table 2 **Expansion of Gulf Merchant Marines**

Country	1976	1980
Bahrain	15	37
Iran	135	208
Iraq	56	123
Kuwait	172	270
Qatar	6	33
Saudi Arabia	55	172
UAE	60	111

Source: Lloyd's Register

Today a large percentage of tankers are registered under flags of convenience. The man who made Liberia a prominent register was Edward Stettinius, former secretary of state under Roosevelt and Truman and the first U.S. delegate to the United Nations. As a good anti-imperialist Democrat, he became concerned with developing Liberia into a maritime country; shortly afterwards Liberia became a flag of convenience.

Title 35 of the Liberian Code of Laws conveniently states, "Income derived by a Liberian corporation owning a vessel registered in Liberia...is exempt from Liberian income tax." One American shipping person estimated that the savings in operating costs on a tanker of 80,000 tonnes amounted to one million dollars a year when it registered under a flag of convenience instead of under the U.S. flag.

In 1970 the eighteen hundred vessels registered in Liberia had never even dropped anchor in a Liberian port. All that was needed was an application form, forty-eight hours to complete the transaction, and a fee of $1.20 per net ton plus an annual tax of 10 cents per ton. It is not surprising that from less than one million tons in 1950, in twenty years the Liberian-registered fleet had swollen to almost thirty-five million gross tons, about 16 percent of the world's total.

Nineteen million tons of that were from oil tankers. Roughly half the Liberian tanker fleet was owned by Greeks, while the rest was owned by Americans, including such corporations such as Exxon (Esso), Gulf Oil, and Texaco. It is significant to point out that since tankers are the largest vessels around, when they run aground and spill, oil a disaster ensues; in this regard one notices that the casualty record of open registers is worse than the world average.

In 1974, Tanker Advisory Service (TAS), an independent fact-finding organization based in New York, produced an analysis of tanker losses over the preceding ten years and noted that the seas were becoming increasingly unsafe. During the decade, 124 tankers had been lost by fifteen national tanker fleets averaging more than one million tons of which exactly half had been flying either the Panamanian or Liberian flag.

For Liberia, the losses represented 0.57 percent of its total tonnage; for Panama it was 0.53 percent. Losses for the more traditional maritime nations were as follows:

- Denmark and the USSR   –   None

- Japan   –   0.03%

- Germany   –   0.08%

- France   –   0.10%

- Britain   –   0.13%

- United States   –   0.17%

- Sweden   –   0.28%

- Norway   –   0.40%

- Greece   –   0.52%

A survey of forty serious tanker accidents by Shell (not of its own tankers) in which a major oil spill occurred indicated that a frequent cause was that "people made silly mistakes." Such mistakes were most likely to occur on ships with crews that were underqualified, improperly trained, and over-worked—that is, on ships flying flags of convenience that had less rigorous manning standards.

In 1972 open register merchant fleets accounted for about 37 percent of the world's shipping accidents; between 1968 and 1972 they accounted for, on average, 33 percent of world shipping casualties. That notwithstanding, flag of convenience operators often say that their vessels, especially many of those under the Liberian flag, are among the largest, best equipped, and most modern in the world. This may very well be true, but the ships are only as good as the crews that run them, and the record of ships under the Liberian flag is not good.

In the case of the *Torrey Canyon*, which was crewed by well-qualified men operating one of the best-equipped ships, nevertheless ran aground off the Sicily Isles because their judgment and seamanship was impaired by terms of service that would not be tolerated on any ship flying, for example, a British or American flag. The Italian master of the *Torrey Canyon*, who had an outstanding record as a seaman, had already served 366 days on board.

Liberian casualties between 1966 and 1970 not only averaged twice as high as those of the major maritime nations, but contrary to the rule that states that old ships have a higher casualty rate than new ones, the Liberian ships on the whole were new ones, certainly newer than the ones lost by the other merchant marine.

In the the case of the *Arrow*, the three-man committee of inquiry led by Dr. P. D. Mc Taggart, executive director of the Science Council of Canada, found that the vessel had been operating without serviceable navigation equipment. The echo sounder had not been working for two months, the gyrocompass had a permanent error of three degrees west, and the radar had ceased to function an hour before the ship ran aground. Furthermore the officer on watch at the time of the accident, the ship's third officer, had no license, the rest of the crew did not have navigational skills, except for the master, and the commission of inquiry had serious "doubts about his ability."

In the commission's final report it stated

> We are well aware of the fact that no form of transportation can be 100 percent safe, but from the record available to us the standard of operation of the world's tanker fleets, particularly those under flags of convenience, is so appalling and so far from this kind of safety which science, engineering, and technology can bring to those who care, that the people of the world should demand immediate action.

In light of this report it is important to delve deeper into the grounding of the *Arrow*.

In December 1998 the Liberian open register had 1,507 ships totaling 49,722,615 tons, or 12.3 percent of the world capacity. An example of the intricacies and problems involved in open registry can be illustrated by the 11,000 ton Liberian tanker the *Arrow*, which ran aground on Cerberus Rock in the middle of Chedabucto Bay, Nova Scotia. The Canadian government set up a task force (Operation Oil) to clean up the mess at a cost of four million dollars. The bone of contention in this case was in trying to recover the cost of Operation Oil from the *Arrow*'s registered owners, Sunstone Marine of Panama.

From the onset one notes that the ship was flying a Liberian flag and was owned by a Panamanian company. This is not unusual, as it was the practice of most bulk cargo carriers to hide the real ownership of the fleet in a thicket of separate corporations; this enabled the ship owner (Onassis) to have it both ways:

1. When he was raising money, the total number of ships controlled by his various companies was acceptable as collateral.

2. At the same time, the assets of each company were limited to one or two vessels.

This ensured that Onassis' empire would never be seriously threatened by the occasional calamity that is an occupational hazard of owning ships. To further complicate matters, the *Arrow* had been operated by Olympic Maritime of Monte Carlo on behalf of Sunstone Marine; she had been chartered to Standard Tankers of Bahamas, who in turn had chartered her to Imperial Oil of Canada.

To get the sort of proof of ownership that would stand up in court, the lawyers had to start by discovering who owned Sunstone Marine, but the lawyers could get no closer to discovering the true owners behind the Post Office Box Corporation in Panama. Sunstone's bearer shares were held anonymously, although the stock certificates were actually in Onassis' possession at the time of the accident. However, even if the lawyers were successful, it would have been a Pyrrhic victory, for the *Arrow* was Sunstone's only asset when it sank—there was nothing left to seize in lieu of damages.

The Canadian government finally gave up proving Onassis was the owner and turned to the Tanker Owners Voluntary Agreement Concerning Liability for Oil Pollution (TOVALOP) to what amounted to an unsatisfied judgment fund. The principal aim of TOVALOP is to provide a fund from which compensation can be paid to those governments who need to take preventive measures in the event of an oil spill and subsequent pollution. TOVALOP funds include provisions for paying the cost of measures to reduce the threat of pollution, even if no spillage actually occurs.

The maximum amount TOVALOP could hand out to the Canadians fell well short of the total cost of Operation Oil. The Canadian government eventually received only $950,000, leaving the Canadian taxpayers to pay the balance of $3 million. The case of the *Arrow* illustrates the ease with which ownership of a vessel can be hidden from a government by using lax laws in different countries to avoid responsibility. The Canadians were not the first to discover the difficulties of finding someone to shoulder the responsibility of a maritime accident by a vessel flying a flag of convenience. The British and the French had gone through similar intricacies in trying to find someone to hold responsible for the grounding of the *Torrey Canyon*.

The *Torrey Canyon* was owned by the Barracuda Tanker Corporation, a financial offshoot of the Union Oil Company of California, which had leased the ship, and had subleased it to British Petroleum Trading Limited which was a subsidiary of the British Petroleum Company. The vessel was registered in Liberia, insured in London, and crewed by Italians. It would appear that finding someone to take responsibility for the accident would be futile.

However, what the British and French did was bide their time until one of the *Torrey Canyon*'s sister ships, the *Lake Palourde*, ambled into a port where the law was believed to be firm. They then seized the ship until the insurers, the only accessible body with responsibility, paid $7.5 million as a settlement for damage.

From 1967 when the *Torrey Canyon* went aground, the next six years witnessed calamities that involved Liberian ships:, the *Ocean Eagle* that spewed oil off the coast of San Juan, Puerto Rico, in 1967; the *Arrow* in 1970; and the *Juliana* in 1971 (which broke in two after hitting a breakwater off the port of Niigata). Two vessels flying the Liberian flag—the Chinese-owned *Pacific Glory* and the Greek-owned *Allegro*—between them carrying 170,000 tons of crude, ran into each other off the Isle of Wright. At the time it was the worst maritime collision on record.

In August 1972 an even bigger collision occurred when two Liberian-flag supertankers—the American-owned *Oswego Guardian*, which was fully laden, and the Greek-owned *Texanita*—collided. The *Texanita*, which was empty, exploded, broke in two, and vanished within minutes. This collision happened on the Indian Ocean northeast of Cape Town; however, it is important to note that half the ship collisions in the world take place in the area bounded by the Elbe River and the English Channel.

For example, between October 1970 and April 1971 ten tankers carrying 3 million tons of crude oil between them were involved in serious maritime accidents in the area; half of them were Liberian, and they included the *Pacific Glory* and the *Allegro* mentioned earlier.

Substandard training of crews and poor maintenance of equipment have been some of the main reasons of these accidents. For instance, on March 3, 1971, the Liberian tanker *Trinity Navigator* ran aground off Barry Head, Torquay, while loaded with 32,999 tons of oil. It was later refloated and it was found that the vessel did not have a VHF radio and its radar was out of order. Furthermore, the Chinese crew spoke no English, the international language of the sea as much as it is of the air.

On April 4, 1971, another Liberian tanker—the *Panther*—was grounded on the Goodwin Sands. Its radar was found to be defective too. The danger such vessels pose to international shipping was one of the principal factors behind the introduction of two-lane traffic in sixty-six busy maritime areas throughout the world. Ships are now required to move in these double lanes of one-way traffic when sailing through these areas, which include the Cape of Good Hope, the English Channel, the Malacca Straits, the San Francisco and New York harbors, the Baltic, and the Straits of Gibraltar. It was felt that this system would reduce the risks to tankers. Unfortunately, many ships ignore these lanes, with disastrous results.

Having discussed the intricacies of open registrars at some length, it is important to look at the actual procedures involved in the registration of ships. To do this we shall look at the British model, as it is one of the oldest established systems in the world and has been adopted by many nations.

International law demands that ownership of any oceangoing vessel is clearly stated and that clear title to the vessel is established. It was the Navigation Acts of 1660 that established compulsory registration of British vessels. Initially, the reasons were not to regulate standards but to ensure that certain trades were reserved for the benefit of British citizens. Generally speaking, all national registrars require that ships under that flag should benefit the owners, who are nationals of the same country, although this can be interpreted more liberally in some countries than in others.

In the case of the British flag, the vessel must be owned wholly by British subjects, although these can be either individuals or corporations.

The actual Certificate of Registry is kept on board the vessel; the full details are also kept in the register itself, normally at the home port of the vessel. This certificate is required by most customs officials. It shows the registered tonnage (GRT and NRT) and is required by most customs officials for entering and clearing ports. Each vessel has its own home port, and the name of that port must appear on the stern of the ship under the name of the ship itself.

Under the British system, British ships must be divided into sixty-four shares and no one really knows why this is so; the sixty-four can either be individuals, corporations, or a mixture of the two. However, ships today are predominantly owned by companies, and this rule dating from 1823 may no longer have much relevance. When a vessel is first registered, the main details will be shown in a register book, including the name of the ship, its port of registry, details of its build and measurement tonnage (the latter obtained from the surveyor's report). Ownership details, including the name and description of the registered owner, together with the proportional shares in the vessel (if there is more than one owner) will also be included in the register book. Any amendments to the shareholding will also be shown.

If the vessel itself is sold, then the relevant bill of sale has to be entered in the book, together with details of the new ownership; such a bill will show any mortgages on the ship. This enables any buyer of the vessel to find evidence of borrowings against the ship when they buy it (i.e., free of encumbrances).

The surveyor's report mentioned earlier states other details that describe the vessel. For instance, the official number of the vessel and the registered tonnage are normally cut into the plate of the vessel and are likely to be found on the after camping of one of the hatches.

Furthermore, a builder's certificate must also be produced for a new vessel (or a bill of sale if a vessel is purchased secondhand). Once the registrar is satisfied with the documents produced, he will then issue some sort of authority (in Britain it is called a carving note), which allows all the approved details to be carved on the vessel, so to speak. After this work is done, the appointed surveyor (who may be appointed by the Transport Ministry or by the classification society for the vessel) will confirm this by signing the carving note and then return it to the ship–owner, who will in turn pass it to the registrar, after which the Certificate of Registry is issued.

The Certificate of Registry is prima facie evidence of title to ownership provided that it is valid, the certificate demonstrates that the ship is entitled to all the rights, privileges and, duties of her flag, and that the last master named thereon is lawfully appointed. Ideally, it is this master who should retain custody of this document. If the vessel should be lost or sold, the Certificate of Registry must be surrendered to the registrar unless, of course, it was lost with the vessel. If this is the case the registrar must be informed of the loss.

4

THE CARGO

In discussing the transportation of crude oil it is in order to look briefly at the formation of the commodity itself. Oil and gas form from algae, bacteria, and other plant matter that is compressed underground in the pore spaces of sedimentary rock. These hydrocarbons are buried under extreme pressure, which increases the temperature of the source rocks. At a given temperature the organic matter transforms itself into oil and gas.

Oil and gas are not necessarily formed together in the same area, as is the case in the North Sea, where the southern part off the Coast of Norfolk mainly consists of gas fields and the northern half of the area is confined almost entirely to oil. Crude oil—the primary concern herein—is not a single entity; it consists of many different parts. It is usually a very dark brown or black, and it is possible to detect yellow or green shades in the oil itself. It also varies considerably with regard to both specific gravity and viscosity.

Many people not involved in the transportation of oil assume that this is a simple exercise whereby one pours oil into the ship at one end and pumps out at the other end. However, every crude oil has slightly different characteristics and must be handled accordingly. It is not surprising, therefore, to find that many problems that arise in tanker chartering are caused by the difficulties in measuring the oil itself. Furthermore, one has to take in further complications that arise in the tankers themselves. Every ship built has its own peculiarities. This also adds to the possibilities of miscalculation and error during ocean voyages. If one also includes the human factor in the carriage of oil by sea, more difficulties can be envisioned.

How does one measure oil? First, oil is lighter than water and therefore has a specific gravity of less than 1.00. However, there are various scales for measuring specific gravity (SG), and the table below shows a number of equivalents between specific gravity (SG) and the American method of measurement called API (for the America Petroleum Institute). This table portrays a general idea of the relation between the various methods of measuring oil cubics. The equivalent API number for a specific gravity of 1.00 is 10, which is the specific gravity of water in the API method of measurement. As the specific gravity drops below the figure of 1.00, the API gravity number rises. Some very heavy crude oils have an API gravity of very close to 10, whereas those of a much lighter variety may well rise as high as the middle 70s.

TALBE OF CUBIC CONVERSIONS

API	SPECIFIC GRAVITY	AMERICAN 42-GALLON BARRELS	CUBIC METRES	CUBIC FEET
10	1.000	6.404	1.018	35.954
15	0.9659	6.630	1.054	37.224
20	0.9340	6.857	1.096	38.497
25	0.9042	7.084	1.126	39.770
30	0.8762	7.310	1.162	42.040
35	0.8498	7.537	1.198	42.315
40	0.8251	7.763	1.234	43.585
45	0.8017	7.990	1.270	44.861
50	0.7796	8.218	1.306	46.136
55	0.7587	8.444	1.342	47.407
60	0.7389	8.670	1.378	48.679

When one looks at the different types of crudes that are shipped around the world it is clear that a wide variation in specific gravity exists, varying from as low as 14.2 API for Bachaquero loaded at Puerto Miranda, Venezuela, to as high as 41.8 for Qatar crude loaded at Umm Said. Further complications arise when one takes into consideration the percentage weight of sulfur. For instance, the sulfur in Eocene crude loaded in the Neutral Zone at Mena Saud is 4.45 percent, whereas that in the Qatar crude loaded at Umm Said is only 1.05 percent.

In general, it can be said that the specific gravity of crude oil varies from about API 12 down to API 45, although the majority of the well-known crudes come within 27 to 35 API. Some crudes loaded from Venezuela such as the Bachaquero crude, though not actually bitumen, have API gravity as low as 14.2, as noted earlier. Later I will show how these various specifications of crude affect the loading and carriage of oil by tankers.

The following, though a very small sample of some world crudes with their API gravities, country of origin, and normal loading terminal, nevertheless helps in the examination and analysis of cargo variations that affect shipping.

TABLE OF WORLD CRUDES

PORT	COUNTRY	CRUDE	API GRAVITY
Ardjuna	Indonesia	Ardjuna	35.20
Arzew	Algeria	Saharan blend	45.50
Arzen	Algeria	Condensate	65.20
Banias	Iraq	Imeg A	36.10
Banias	Iraq	Imeg B	35.10
Barrow Island	Australia	Barrow Island	36.60
Bejala	Algeria	Hassi Messaoud	43.00
Bombay	India	Bombay High	39.40
Bonny	Nigeria	Light	36.70
Cabinda	Angola	Cabinda	31.70
Ceyhan	Iraq	Kirkuk	35.10
Dairen	China	Taching	33.00
Das Island	Abu Dhabi	Umm Shaif	37.46
Das Islanc	Abu Dhabi	Zahum	40.60
Dubai	Dubai	Dubai Export	33.50
Dumai	Indonesia	Sumatran Light	34.50
Es Sider	Libya	Es Sider	36.76
Forcados	Nigeria	Forcados Blend	29.70
Hound Point	U.K.	Forties	36.60
Jebel Dhanna	Abu Dhabi	Murban	40.50
Kharg Island	Iran	Light	33.80
Kharg Island	Iran	Heavy	31.00
Kole	Cameroon	Kole	34.90
La Salina	Venezuela	Tijuana Light	32.10
La Skhirra	Algeria	Zarzaitine	43.00
Long Island	Australia	Gippsland	46.40
Lutong	Brunei	Seria light	36.20
Marsa El Brega	Libya	Brega	40.40
Marsa El Hariga	Libya	Saris	38.30

TABLE OF WORLD CRUDES (Continued)

Mena Al Ahmadi	Kuwait	Kuwait	31.40
Mena Al Fahal	Oman	Oman Blenn	36.30
Mena Saud	Neutral Zone	Eocene	16.50
Pajaritos	Mexico	Isthmus	32.80
Pajaritos	Mexico	Maya	22.00
Pennington	Nigeria	Pennington	36.60
Puerto Miranda	Venezuela	Bachaquero	14.20
Qua Iboe	Nigeria	Qua iboe	35.80
Ras al Khafji	Neutral Zone	Khafji	28.60
Ras Lanuf	Libya	Light	37.80
Ras Tanura	Saudi Arabia	Light	33.40
Ras Tanura	Saudi Arabia	Medium	30.80
Ras Tanura	Saudi Arabia	Heavy	27.90
Statfjord Buog	Norway	Statfjord	38.40
Sullom Voe	U.K.	Brent Blend	38.00
Tess	Norway	Ekofish	43.40
Tripoli (Lebanon)	Iraq	Imeg A	36.10
Tsingtao	China	Shengli	24.20
Umm Said	Qatar	Qatar	41.80
Valdez	US	Alastia North Slope	26.40
Zueitina	Libya	Zueitina	43.00

Let us look now in more detail at where the various types of crude oil can be found and where they are loaded, starting with the Middle East. The term applies to the region that includes Southwest Asia and part of northeast Africa, lying west of Afghanistan, Pakistan, and India.

This region contains much of the world's oil reserves and has many strategic trade routes (e.g., the Suez Canal). The oil industry dominates the economics of the countries in the Middle East, with Saudi Arabia being the largest producer of crude oil in the world. The main loading terminals in Saudi Arabia are Ras Tanura and Juaymah. At the end of 1986 Saudi Arabia had an estimated reserve of 166.6 billion barrels. This figure represents nearly 24 percent of the entire reserves of oil throughout the world. At the end of 1988 the Saudis upgraded their reserves by 52 percent to 252.4 billion barrels. Other reappraisals by OPEC (Organization of Petroleum Exporting Countries) states, particularly Iran, Iraq, Abu Dhabi, Dubai, and Venezuela boosted the estimated OPEC reserves since 1986 by around a whopping 60 percent.

Iraqi oil, which is of a medium type and varies in specific gravity from about 31 to 36 API, used to be shipped out of Iraqi ports in the Arabian Gulf. However, due to recent wars, very little oil is being exported. Traditionally,

Iraqi crudes have been moved by pipeline through to Banias in Syria, the Lebanese city of Tripoli, and Ceyham (Botas) in Mediterranean Turkey.

Iranian oil fields are largely in the northern part of the country and consist of light and heavy oils. Consequently, Khorg Island has become the predominant loading terminal. The Iranians also ship some of their oil, which varies between API 31 for heavy and 34 for the light, from Sirri Island and through a shuttle service down to Laven (Hormuz) Island.

Moving to Kuwait, which exports its oil out of Mena Al Ahmadi and through the Neutral Zone (which it jointly owns with Saudi Arabia) to the loading terminals at Mena Saud and Ras Al Khafji. Bahrain is not a major exporter of crude, as it is mainly involved in refining. Qatar, on the other hand, exports its crude through Umm Said, Abu Dhabi through Das Island and Jebel dhana; Dubai exports through Dubai itself and also through Feteh. Crudes from the state of Oman are loaded at Mena Al Fahad.

Pipelines were mentioned briefly in regard to crude oil from Iraq. It is fundamental to note the impact future pipeline development in the desert between the Arabian Gulf, the Red Sea, and the Mediterranean will have on the tanker trade. Any oil shipped out of the Red Sea to Western Europe and the United States through the Suez Canal involves a much shorter sea passage than oil coming directly from the Arabian Gulf; any reduction in sea passage miles will affect the world's tanker fleet leading to a possible oversupply of tonnage and a reduction in rates.

Since September 11, 2001, the world's focus has been shifted toward combating terrorism. Pipelines are particularly vulnerable to terrorist attacks and therefore it may be safe to speculate that few if any major pipelines will be built in the near future. That notwithstanding, it is appropriate to note that the most important pipeline in the region is the petroline that runs across Saudi Arabia to Yanbu on the eastern shore of the Red Sea.

While the Middle East is without a doubt the dominant area for the supply of world crude, there are many other countries that export oil in large quantities. In terms of actual crude reserves, there is little doubt that Africa comes high on the list with reserves distributed among a number of countries such as Nigeria, Libya, Gabon, Algeria, Angola, Egypt, and Tunisia. Due to the political instability of the Gulf region Africa is increasingly becoming an important source of oil for the U.S. economy.

Nigeria is the dominant exporter of crude from West Africa. Cargoes are shipped out from Bonny, Forcados, Pennington, Qua Iboe, Brass River, and Escravos. Lesser quantities are exported by other countries such as Cameroon

(Kole Terminal), Gabon (Cape Lopez), Congo Republic (Djeno), and Angola (Cabinda). In northeast Africa, Libya is the predominant producer of crude. Libya is not the dominant force it used to be with regard to the actual market transportation of crude. This is due to the United Nations sanctions that were imposed on the country after the downing of Pan Am Flight 103 over Locker-bie, Scotland, in 1989. Nevertheless, at the time of this writing Libya is trans-forming itself from a country that condones terrorism to one that rejects it. Sanctions were lifted and Libya is likely to play a greater role in the transpor-tation of oil. This, however, does not mean that Libya is not continuing to produce in appreciable quantities. The main loading terminals in Libya are Marsa El Brega, Es Sider, Zueitina, Ras Lanuf, and Marsa El Hariga.

Egypt is another important African exporter of crude, particularly since the summed pipeline runs across the country parallel with the Suez Canal. Although Egypt can be considered a Mediterranean country, it in fact ships all its oil from ports in the Red Sea south of Suez, such as Ras Gharib, Ras Shukheir, Ras Budran, Wadi Feiran, and Zeit Bay.

As there are limitations on the Suez Canal for the size of laden vessels that can transit, it has therefore proved prudent to use the summed pipeline by car-rying oil form the Gulf to Ain Sukhun the terminal at the southern (Red Sea) and at the northern or Mediterranean terminal of Sidi Kerir. At Sidi Kerir smaller vessels are usually able to distribute this oil to the various ports in Western Europe and the United States. The only important oil exporter in the Mediterranean is Algeria, and the terminals used are Arzew Bajaia, La Skir-rha, and Skikda.

The continent of South America extends from Chile in the south to Colombia in the north. Mexico (central America) and Venezuela are impor-tant producers of crude. Although Mexico holds the largest reserves on the continent, As far as the world's tanker fleet is concerned, Venezuela is far more dominant in the export of crude than Mexico, although quite a large proportion of Mexican crude goes through the Iberian Peninsula.

Venezuela has a large number of terminals along its coast, particularly in the Lake Maracaibo area, including Punta de Palmas, Bajo Grande, La Salina and Puerto Miranda (at the mouth of the lake), Amuay Bay, Punta Cardon, Caripito, Puerto La Cruz (all on the north coast) and Puerto Orduz (up the Orinoco River). Oil exports from Mexico come from the east coast and the most important ports are Pajaritos, Cayo, Arcas, and Dos Bocas.

When one looks at the Pacific, the only real reserves on the west coast of North America are the oil fields in Alaska, where the main export terminal is

Valdez. In the southern hemisphere exports can be found from Ecuador (Balao and Esmeraldas) and Peru (Callao and Talora). Crude has been found relatively recently in Australia and is shipped in what can be considered small quantities, as compared to other countries, through Barrow Island, Jabiru, and Botany Bay terminals. Moving north from Australia one comes to Indonesia.

Indonesia is important to world shipping in that a large proportion of their production is exported even though proven reserves are small. The terminals capable of handling large vessels for export include Balikpapan (Bornes), Ardjum, Balongin, and Cinta (Java), and Dumai (Sumatra). Other exporter countries in the region include Malaysia, which ships its crude out of ports such as Mirri, Labuan, and Bintulu. Brunei's oil is loaded at Seria. The growing importance of China will make the country an active participant in the exporting of crude, although their crudes can be very heavy. The Chinese terminals include Weizhou, Dairen, and Tsingtao.

Three areas of the world that need to be looked at closely are northwest Europe, the internal production in the Untied States, and the nations of the former Soviet Union. Production in the United States is not of much relevance as most of their production is consumed within the country, whereas Western European exports are extremely important to the tanker chartering market. North Sea reserves owned both by the United Kingdom and Norway are relatively small in comparison with other areas, but in the shipping world the North Sea is important in tanker operations.

The most economical way of running a vessel is to limit the ballast voyage to as short a distance as possible, but at one time there was absolutely no alternative for a VLCC discharging in Western Europe other than to proceed in ballast back to the Arabian Gulf or, at the very best, West Africa. Nowadays it is possible for such a VLCC, after discharging on the continent, to load crude out of the North Sea and proceed to ports on the United States' east coast.

The United States is important as a net importer of oil, and this greatly affects the free tanker chartering market. The former Soviet Union was at one time the largest producer of crude oil, and today Russia and other former Soviet Republics have massive reserves of oil. However, their impact on world shipping has been small. Nevertheless Black Sea ports such as Batum and Novorossisk are important oil terminals.

General Characteristics of Crude Oil

Anyone interested in the transportation of crude will need to have some basic knowledge of crude oil. Crude oil consists of many parts and some of the constituent parts will have a greater or lesser effect on the oil tankers themselves. The main element is carbon, which accounts for about 85 percent, followed by hydrogen, which makes up most of the remaining 15 percent, leaving a small percentage of sulfur (up to about 6 percent), and about 1 percent each of oxygen and nitrogen. What remains are small amounts of trace elements, metallic and otherwise. In the transportation of crude these various ingredients are not of much importance, except perhaps sulfur and oxygen, which can increase the risk of tank and line corrosion.

That notwithstanding, it is important to emphasize that the hydrocarbon elements of crude oil contain components such as wax, asphalt, sand, and water, and it is these latter characteristics that can cause problems for the vessel. Additionally, the light ends of crude oil are liable to evaporation, and these are the more valuable parts of crude. Tanker operators try to minimize such losses; loss of valuable light ends in transit is more directly of concern to the owners of the cargo and can result in a shortage of outturn, and this would obviously involve the ship owner.

One may quite correctly ask how evaporation can take place in the cargo tanks, which are for all intents and purposes not only enclosed but also inverted. True, in a well-maintained vessel there should only be a limited amount of loss while at sea, but it is difficult to ensure that a vessel sailing through heavy seas is totally gas tight. It is important to point out, however, that most losses occur during loading and discharging. Obviously, such losses should be minimal in a well-maintained tanker, but some loss of cargo during a tanker voyage is inevitable.

It goes without saying that the potential for these losses will be increased if the crude needs some form of heating, and this brings up other matters of interest in oil transportation, such as pour point, cloud point, and viscosity.

Pour Point

The lowest temperature at which oil is observed to flow is its pour point, and this is measured by reducing the oil temperature gradually until it causes flow. Pour points of crude oils vary considerably depending on the wax content of the crude. The percentage of wax in individual crudes brings us to the second important factor:

Cloud Point

This point is the temperature at which the waxes in the liquid begin to visibly change into solids. While such a charge is visible in transparent liquids, including distillate levels, this is not the case with crudes, as they are usually too opaque. In such oils, the cloud point is usually obtained by measuring the change in viscosity.

Viscosity

Viscosity is defined as the internal resistance a liquid has to flowing under gravity. Oil that flows freely is said to have a low viscosity, while thick and sluggish oils have a high viscosity. Basically, viscosity is measured by the speed with which an oil passes through a given length and diameter of pipe and, of course, at a standard temperature. Under the Redwood system, oils are measured in seconds and this gives a typical intermediate fuel oil a reading of 1,500 seconds and a heavy fuel oil, as used in ships' boilers, of 3,500 seconds. However, the more modern method is to measure viscosity in centistrokes (CST), and the equivalents for the fuels we have just mentioned would be 180 CST and 380 CST.

These three factors are of direct interest to tanker operations, particularly when one considers the measurement of oil. Tanker owners try to categorize the crudes they prefer so that the life of the ship can be prolonged and maintenance kept at a minimum. Below are a few sample figures of some general crudes. With this it is possible to see how they vary and in which way this variation can affect a tanker.

CRUDE	API	POUR POINT	CLOUND POINT
Arabian Light (Saudi Arabia)	33.4	−32°C	+27°C
Ardjuna (Indonesia)	35.2	+25°C	+31C
Brass River (Nigeria)	40.1	+1°C	+25°C
Brent (North Sea)	38.0	−11°C	+27°C
Bu-Attifel (Libya)	43.6	+37°C	+51°C
Chinese	24.2	+34°C	+48°C
Gulf of Suez (Egypt)	31.9	+13°C	+27°C
Iranian Heng (Khorg Island)	31.0	−12°C	+27°C

From the selected crudes, we can see that these vary from a low of 24.2°C for Chinese to 43.6°C for Bu-Attifel (loaded at Zietina). Storage factor is affected by API and thus it is possible to load a higher quantity of Chinese crude than Bu-Attifel in a given space. This fact is of considerable importance to a ship owner and master since:

1. The earnings are based on the deadweight quantity loaded.

2. The stowage factor will have an effect on any segregation of grades and the trim of the level.

The difference in stowage between these two samples is approximately 7.05/7.93 bbls/ton, or 12 percent. An average tanker with a capacity of 630,000 bbls and a deadweight cargo capacity of 89,000 long tons could load a full cargo of Chinese crude, but would be limited, with Bu-Attifel, to about 79,450 long tons and a consequent loss of freight.

When considering the issue of storage, it is important to note that most tankers now have SBTs (Segregated Ballast Tanks) and this anti-pollution requirement has considerably reduced the cubic capacity per cargo. Thus, stowage factors for oil are even more important than they formerly were.

The second factor to consider is the Pour Point. The crude with the most attractive (lowest) stowage factor, Chinese, also has a relatively high Pour Point of 34°C (93.2°F). It is apparent that more degrees of heat will be required before this cargo can be easily pumped into and out of the vessel. Bu-Attifel is even less attractive, as its Pour Point is even higher at 37°C (98.6°F). Also note that Brent and Iranian Heavy have a minus figure and will obviously be easier to handle at lower temperatures.

Finally, there is the issue of Cloud Point. Tanker operators do not under any circumstances want waxes forming in the cargo tanks. Once solidified, it cannot be changed back to a liquid merely by returning the temperature to above the cloud point. Furthermore, the wax not only forms on the cargo tank sides and bottom but also on the heating coils, reducing the effectiveness of the coils by acting as a partial insulation. Once there is substantial build-up of wax in the cargo tanks, the only way to remove it is to dig it out. Just as a high Pour Point will produce difficulties in cargo handling, so will a high Cloud Point. This makes the Chinese crude oil and even more so, the Bu-Attifel very unattractive cargoes.

However, when dealing with market forces, we may have little choice regarding the cargoes carried, but this may not always be the case, and awareness of potential problems will make them easier to avoid. In the long run, the life of a tanker can be prolonged, providing a better return on capital. Wax is not the only constituent of crude that can cause trouble; sulfur can cause some damage too if present in too high a proportion, as it is a corrosive substance. However, this is less likely in crude, but may cause damage to internal combustion engines, and is therefore a consideration when it comes to bunkers.

Water

All crudes contain a certain amount of water, and nearly all crudes have a specific gravity of less than 1.00. The oil will initially rise to the top of the cargo hold. This makes it easier to separate the two liquids and allow the ship to decant the free water. However, it is also important to differentiate between this free water that settles out and the water that remains in solution with the oil. This forms part of what is known as base, sediment, and water (BS&W), and once again these impurities will be found to a greater or lesser extent in all crude oils. These items add to the problems of oil measurements.

As noted above the characteristics of a cargo of crude can affect the transportation by sea. The most important characteristic is the specific gravity and therefore, its stowage factor. It is crucial to know the approximate stowage factor of the cargo to ensure that the vessel can carry a full load and earn freight accordingly. In large tankers, such as the VLCC and ULCC, the problems of storage factor do not arise very frequently, but nevertheless it is a consideration.

For instance, consider a typical VLCC with a deadweight cargo capacity of about 260,000 long tons and a cubic capacity of 11.6 million cubic feet. This

calculates out to very close to forty-five cubic feet per ton. Consulting a specific gravity table, this enables such a vessel to load a cargo of gravity up to
about 45 API, or 0.8 specific gravity. As this specific gravity covers virtually all
the crudes loaded out of the Arabian Gulf, there is usually no great problem
with the large vessels.

One must also bear in mind that specific gravities, as given in tables, are
always calculated at a temperature of 60°F and, should it be necessary to have
a cargo loaded at a higher temperature than this, then the specific gravity will
decrease, resulting in a higher stowage factor. This problem hardly ever arises
with large vessels, most of which are not fitted with heating coils and are
unable to carry cargos that require heating.

When, however, we come to look at a medium size tanker of approximately
110,000 deadweight with SBT/CBT, this may well have cargo with a capacity
of only about 3.8 million cubic feet. Assuming her deadweight cargo carrying
capacity is about 108,000 tonnes, this only calculates out to about thirty-five
cubic feet to the ton available per cargo, if the specific gravity of the oil was
very close to that of water at 1.00 or 10 API.

Refined or Clean Products

Clean products that come from crude can be divided into two categories:
leaded and unleaded. It does not make much difference to the ship itself
whether the cargo contains lead, but it is of vital importance to the cargo owners. However clean a ship's tanks are prior to loading products, there are
bound to be some residues of previous cargo. If this residue is lead-based and
becomes mixed with a new lead-free cargo, considerable damage could result.

The more refined the product, the more care needs to be taken with its
transportation. For example, it is important that jet fuel does not acquire
impurities through tank contamination. In fact, most charter party agreements
contain a clause that requires the vessel to pass an inspection by a charterer's
surveyor before commencement of loading.

Clean products carried in tankers can also be divided and defined using the
NPA (for the National Petroleum Association) color scale.

NPA Colors

NPA	No. 1	Lily White	No. 4 Orange Pale
	No. 1½	Cream White	No. 4½ Pale
	No. 2	Extra Pale	No. 5 Light Red
	No. 2½	Extra Lemon Pale	No. 6 Dark Red
	No. 3	Lemon Pale	No. 7 claret Red
	No. 3½	Extra Orange Pale	No. 8 Extra Dark Red

The colors of all oils vary, including products. The above scale grades crude and products by color and runs from Lily White (No.1) to Extra Dark Red (No. 8). Generally speaking, the lighter the color, the more refined the oil, and this has established the dividing line between dirty and clean oils at 2½ NPA (Extra Lemon Pale). For many reasons, it is normal to insert in all charter parties for clean products phrases such as "Clean Petroleum Products unmarked than two and a half NPA." In this leg, the ship owner ensures that his clean product tanker will not be contaminated by the loading of oil that is too dark in color, and therefore almost certainly containing more impurities. Whether or not the actual type of cargo is specified in a charter party, the NPA qualifications should always be added because certain products such as Naphtha and gas oil can be both clean and dirty on the color code. Without the two and a half NPA here, a charterer might load, say, dirty Naphtha and still be within the charter party.

The Paradox of Oil

The exploration of oil reserves does not automatically lead to prosperity. Oil revenues can be used prudently or foolishly, and poor development outcomes are not inevitable in oil-exporting countries, but their performance over the past thirty years seems to suggest oil may be a curse.

Some of the world's exporting countries mentioned earlier (with the notable exception of Norway) are among the most economically and politically troubled. Many of these countries have high poverty rates compared with countries dependent on the export of agricultural products; infant mortality, malnutrition, and life expectancy at birth is worse than in non-oil/mineral dependent countries of the same income level. Their economies tend to grow more slowly than countries that are not oil-rich over time (between 1965– 1980 OPEC members experienced an average decrease in their per capita GNP of 1.3 percent per year, while their non-oil counterparts grew by an average of 2.25 percent per year).

However many countries aspire to joining the ranks of oil-exporting countries. The potential of African oil coming on stream is likely to bring about a fundamental change not only in the domestic economy but also in the sphere of geopolitics. Countries in East Africa have the potential to become exporters of crude in the medium term (within ten to fifteen years).

Kenya, for example, currently has bitumen that can be commercially mined, particularly in the Mandera region where the National Oil Corporation (NOOK) found massive reserves after drilling a shallow well. On July 31, 2003, the Norwegian vessel *Polar Princess* docked in Mombassa and in the following weeks became the exploration base for a team of geologists from Woodside Energy, the firm that is hunting for oil in the deep seas of the Indian Ocean off the coast of Lamu.

Previous efforts to find oil in Kenya have been described as lackluster by oil search experts. Only thirty exploration wells have ever been sunk, compared to Sudan, which found massive oil reserves after sinking seventy-eight wells. Experts contend that prospecting died out in Kenya in 1992, just when it looked certain that a major discovery was about to be made. However, renewed interest in prospecting for oil in Africa is likely to lead to better results.

An interesting revelation of Kenya's crude and natural gas potential can be obtained in the 1993 World Geological Survey, conducted by the United States Department of the Interior. The survey, which attempts to establish reserves of unidentified crude oil and natural gas in the world, found that Kenya's coastal region has potential for producing around one hundred million barrels of crude oil and between six hundred billion and six trillion cubic feet of natural gas.

The report identified the basins around the Kenyan, Somali, and Tanzanian coast as having the same potential for undiscovered crude oil as the Gulf of Guinea, which has now become the global hotspot for oil exploration after discoveries in Equatorial Guinea. The Gulf of Guinea touches Nigeria, Gabon, Cameroon, Equatorial Guinea, and the Democratic Republic of Congo. Interestingly, since that survey, Somalia has discovered it has a huge crude oil potential.

However, as countries dream of oil revenues it is important to note that oil-exporting countries are more likely to spend two to ten times more on their militaries and to be ruled by authoritarian leaders. For example, in November 2000 the Chad government used part of a $25 million oil contract bonus to purchase $4.5 million worth of weapons. President Idris Deby, who came to power by means of a military coup, justified the purchases as being consistent with the demands of national security and the need to protect development. Furthermore, the localities surrounding oil installations are among the most environmentally damaged and conflict-ridden in the world. Worst of all, the probability of having civil wars is higher in oil-exporting countries than in their resource-poor counterparts.

A World Bank study carried out by Paul Collier and Anke Hoeffler attempted to find the links between natural resources (including oil) and conflict. The Collier-Hoeffler model, after testing for a number of factors, found that there are three significant ones:

- The level of income per capita; doubling per capita income roughly halves the risk of a civil war

- Rate of economic growth; each additional percentage point of growth reduces the risk by about 1 percent

- Structure of the economy, namely, dependence on primary commodity exports (e.g., oil)

What is very intriguing about the Collier-Hoeffler model is that they found that the effect of primary commodity dependence is nonlinear, peaking with exports at around 30 percent of gross domestic product (GDP). This means that a country that exports around 25 percent of GDP has a 33 percent risk of conflict, but when such exports are only 10 percent of GDP the risk drops to 11 percent.

Why, then, does dependence on oil seem to produce such perverse development consequences? Why this "paradox of plenty"? The answer is (and the one most people living in oil-producing/exporting nations tend to agree with) is that the oil sector is particularly prone to corruption. Examples abound: in Equatorial Guinea, the U.S. Justice Department has been asked to investigate how $500 million came to be paid into a private U.S. bank account, said to be under the control of the president. In Nigeria a subsidiary of Halliburton admitted that it paid millions of dollars to an official in return for tax breaks. In Angola, more than $1 billion per year of oil revenues disappeared during

1996–2001—fully one-sixth of the nation's income. This is particularly disturbing in that the majority of Angolans—some 70 percent—live on less than a $1 per day. Corruption is not only an African problem. In Kazakhstan Exxon, Mobil, BP, Amoco, and Phillips Petroleum made payments of more than $1 billion to President Nazarbayev and other senior officials.

The history of oil-exporting countries shows that capturing oil profits, legally through rent-seeking or illegally through corruption, has proved to be a very lucrative venture. Leaders seek oil contracts that enrich themselves, even if these contracts lower the overall social welfare of a country. The socioeconomic cost of corruption and widespread rent-seeking is massive. Rulers and senior officials tend to favor capital-intensive mega-projects in defense and infrastructure, in which payoffs can be concealed over more beneficial health and evaluation expenditures that might enhance the quality of social and economic services over time. This policy slows growth and lowers aggregate income.

But corruption explains only part of the oil paradox. It is very difficult for a rent-seeking oil economy to utilize its reserves well, because they are especially susceptible to policy failure. This can be explained by a combination of factors, including:

1. The especially high price volatility of oil (twice as volatile as other primary commodities since 1970);

2. The long-term price deflation of oil compared with the cost of imported manufactured goods, which tends to promote adverse balances of payments

3. The poor employment generation and technical diffusion and transfer of the petroleum sector

4. The Dutch Disease, an economic phenomenon that states that a rise in export earnings leads to an appreciation of the real exchange rate in a boom period and a loss of competitiveness for tradable goods sectors not benefiting from the boom. Thus, oil booms eventually lead to the collapse of productive industry, especially agriculture. This in turn encourages an even greater dependence on oil.

Therefore, it is prudent to ask how an oil economy revives its productive industries. One way is to diversify to reduce the dependence of governments on a single commodity. States with more diverse exports are better protected against international market fluctuations and less prone to the *resource curse* (the inverse association between economic growth and natural resource dependence). For oil exporters, an obvious route to diversification is to develop downstream industries, which can process and add value to raw materials.

This is not an easy proposition for an oil exporter because the advanced industrial states in the Organization for Economic Co-operation and Development (OECD) place higher tariffs on refined oil than on crude to protect their own oil-processing firms against competition; OECD countries place no tariffs at all on the importation of unprocessed oil and on many minerals. The table that follows highlights the tariffs associated with oil.

Mean OECD Tariffs on Processed and Unprocessed Extractive Products
Product and Description Tariff
Petroleum oils, crude oil 0.00
Petroleum resins, coumarone, indene, or coumarone-indene resins, and polyterpenes 7.00
Woven fabrics made from high-tenacity yarn, nylon, or other
Polyamides or polyesters 8.47
Polyethylene (used for grocery bags and children's toys) 6.87
Polymers of vinyl chloride (PVC plastic) 7.52
Polycarbonates (used for light fittings, kitchenware, and compact disks) 7.84
The removal of tariffs and nontariff barriers to value added goods by the OECD states could greatly assist states dependent on oil to diversify.
(Source: UNCTAD-Trains database, available at www.unctad.org)

5

THE RHODESIAN OIL EMBARGO

The previous chapter looked at oil as a commodity vis a vis shipping. How-ever, one cannot ignore the fact that oil has also been used as a political weapon. In this regard the tanker industry has played its role as a facilitator of political motives. Powerful multinationals have influence over the foreign pol-icies of nation-states. What follows is an account of how powerful interna-tional oil companies frustrated the attempt to crush the Ian Smith regime in Rhodesia by sanctions. The following highlights the politics of oil in regard to its transportation.

At the time of this writing the interim report by the Independent Inquiry Committee (IIC) on the Iraq Oil-for-Food Program that ran from 1996 to 2001 was newly released. It is important to state that Saddam Hussein was sanction-busting since 1991, before the Oil-for-Food Program was estab-lished. Where was the oil going? And how did it get where it was going? The major markets for Iraqi oil are in Europe and North America. Inevitably, tankers were involved in sanction busting. One shall have to wait for the final report to see how wide the web of deceit extends and how the oil was trans-ported.

The issue of oil embargoes and sanctions is beyond the scope of this book, but it is still important to highlight how the tanker trade, open registries, and shrewd businessmen come together to compose a formidable force in interna-tional trade and relations. It is not surprising, therefore, to note that the trans-portation of oil is sometimes a political as much as an economic exercise. As the Iraqi Oil-for-Food inquiry unfolds, it is important to look at the Rhodesian case of sanction busting exposed by Martin Bailey in *Oilgate*.

After Prime Minister Ian Smith's Universal Declaration of Independence (UDI) on November 11, 1965, Rhodesia became the first colony to break with Britain without consent since the American Colonies did so in 1776. Britain called Rhodesia's action illegal and banned all trade with Rhodesia. In 1966, the United Nations (UN) imposed economic sanctions against Rhodesia. A new constitution, approved mainly by white voters, was designed to prevent the black African majority from ever gaining control of the government.

The constitution took effect in 1970 and Rhodesia declared itself a Republic on March 2, 1970. No country recognized its independent status, and many countries continued to apply political and economic pressure through the UN to end white minority role in Rhodesia. Rhodesia was landlocked and therefore completely dependent on its neighbors for access to the outside world. South Africa and Mozambique refused to accept the UN decision to impose sanctions, enabling the Smith regime to trade with the outside world.

This was facilitated through a railway line running from Rhodesia to Mozambique's port of Lourenco Marques (renamed Maputo at independence). This route was of particular interest because there was no direct railway link between Rhodesia and South Africa (although a line was under construction). Rhodesia had no oil of its own, yet the country consumed some 15,000 barrels of oil a day. It was no secret that Rhodesia's oil was supplied by rail from Lourenco Marques.

These revelations came to the public through a pamphlet called "The Oil Conspiracy." This document was the work of Bernard Rivers, a freelance researcher on Third World affairs who had been sent to Lourenco Marques by Granada Television as a consultant for a *World in Action* film about Mozambique. In his spare time, he investigated the sanction busters. *The Oil Conspiracy* was launched at a press conference in Washington DC on June 21, 1976, and it identified Mobil subsidiaries supplying Rhodesia with oil. It also showed that arrangements were made to conceal Mobil's involvement in sanction busting. Other oil companies, particularly Shell, were also mentioned.

Oil was undoubtedly very important to the survival of Rhodesia, for oil was not only needed to fuel the economy but also to ensure the mobility of the armed forces that were involved in a guerilla war. To insure that oil was available, Rhodesia's first oil refinery was opened at the border town of Umtali a few months before UDI. This new refinery was fed with crude oil by a pipeline from the port of Beria in Mozambique. There was concern that an international embargo might cut the flow of crude oil to Beria, thereby rendering the Umtali refinery useless.

It is important to note that, apart from the rail link to Beria, Rhodesia also had a rail outlet to the port of Lourenco Marques. When UDI actually came, Mozambique and South Africa remained neutral in the dispute between Britain and Rhodesia; this neutrality meant nonparticipation in imposing the UN embargo. At this juncture it is pertinent to point out that the Security Council resolution merely *recommended* the imposition of sanctions, it was not mandatory.

However, several oil-exporting countries stopped selling crude oil to Rhodesia. Iran, then the main supplier of the Umtali refinery, cut oil shipments on November 22, 1965. Kuwait, whose national oil company had a small shareholding in the Umtali refinery, and Libya also took similar action. International oil companies respected these embargoes, but diverted shipments from elsewhere to the Umtali refinery. A convenient source was the British Protectorate of Abu Dhabi.

In early December, a BP tanker—the *British Security*—steamed out of Abu Dhabi and docked in Beria with eighty thousand barrels of crude oil for Rhodesia on December 14. The actual UK Sanctions Order introducing an oil embargo was issued on December 17, 1965, and its crucial flaw was that it only covered companies registered in Britain, not their subsidiaries, which were incorporated in foreign countries.

On the day the sanctions order was introduced, two tankers were already steaming for Beria. One was the *Staberg* owned by Shell carrying eighty thousand barrels of crude for Rhodesia.

Following in the *Staberg*'s wake was the *Tamarita*, a Norwegian tanker chartered to Aminoil (American Independent Oil Company). The *Staberg* eventually altered course and Aminoil ordered the *Tamarita* to discharge at the Kenyan port of Mombassa.

On January 15, 1966, the Umtali refinery closed down, for no crude oil had been delivered at Beria since the introduction of the sanction order. However, the South African government defeated the oil embargo through a carefully planned strategy. First, by announcing its neutrality. It then allowed the exports of "traditional" specialized oil products. This was followed by "gifts" of nontraditional supplies. Once this traffic had been established, private companies joined the trade.

South Africa initially defied the oil embargo by the road route via Beit Bridge. This was an expensive way of transporting oil so the Rhodesian's turned to the railway that runs from the Mozambican port of Lourenco Marques into Rhodesia. The advantage of dispatching oil from Lourenco Marques is that the local Sonarep refinery was controlled by Portuguese interests and therefore was not under quite the same restraints that the British—and American-owned refineries in South Africa faced in trading with Rhodesia.

The Rand Daily Mail of March 10, 1966, reported that Rhodesia was receiving as much as 4,600 barrels a day of oil products by rail from Mozambique. This quantity represented over half of Rhodesia's normal oil requirements, which with rationing would provide the bulk of the country's fuel needs. It was clear to Britain's leadership that they had to do something to stop the sanction busting.

Sanction busting was a multinational scheme involving shipments of Iranian crude sold by Lebanese subsidiaries of an American company to two Greek firms. The oil was then to be carried in a Greek registered tanker, owned by a Panamanian company, and chartered to a South American firm. This exercise involved a brisk turnover in ships' names, flags, captains, and destinations, and was a dramatic illustration of the difficulties of disentangling the labyrinthine knots of the international tanker business at work in the sanction-busting chain. What follows is the account of the voyage taken around Africa in 1966 by the *Joanna V.* The voyage of this vessel is a very good example of how complicated and political the transportation of oil can be.

The *Joanna V*

On February 21, 1966, the *Arietta Venizelos* sailed from the Iranian port of Bandar Mashur bound for Rotterdam. It was laden with a cargo of 110,000 barrels of crude oil originally belonging to San Jacinto, a Beirut-based subsidiary of U.S. Continental Oil. Nicos Vardinoyannis, a Greek businessman, had bought the oil through two of his companies: Seka and Nima International. At this time Vardinoyannis had made arrangements to charter the tanker from Venizelos—the Greek shipping firm that owned the *Arietta*—to carry the cargo to Beria.

On March8, 1966, Nicos Vardinoyannis attempted, while it was in the Mediterranean, to purchase the vessel from the company that had chartered it to him. His offer of forty thousand pounds (about double its nominal value) was quickly accepted.

On March 12, 1966, arrangements for actual purchase by a Panamanian subsidiary, Varnikos Corporation, owned by Vardinoyannis, was completed in New York with a letter of credit from Johannesburg. Meanwhile efforts were being made at the port of Beria to receive the oil already on its way in the *Arietta Venizelos*. However, one obstacle existed in that it was not possible to pump the oil directly from the tanker into the Umtali pipeline. The oil had to go through the storage tanks belonging to an international oil company before dispatch to Rhodesia. These international oil companies had, however, made it clear that they would not cooperate in supplying oil to Rhodesia.

On the high seas under its new owner, the *Arietta Venizelos* was chartered to the Cape Town firm of AG Morrison and was ordered southward to the Senegalese port of Dakar, where the younger brother of the new owner flew out from Athens to captain the tanker to Southern Africa. The tanker was now renamed the *Joanna V* and, since the vessel had been registered at Piraeus, the Greek maritime authorities had warned George Vardinoyannis via radio that it was illegal for a Greek vessel to deliver oil to Beria for Rhodesia. Nevertheless, by the end of March the *Joanna V* had rounded the Cape and was approaching the Mozambique Channel.

The tanker was intercepted by the British frigate *Plymouth* on April 4. The captain explained that his instructions were to put into Beria for bunkering and provisions and then proceed north to Djibouti to discharge his cargo. This was hogwash, for Djibouti had no oil refinery.

The Royal Navy could not enforce the embargo by preventing the tanker sailing for Beria. This could only be done with the prior approval of the government that had registered the vessel, and the Greek Foreign Ministry was willing to give such authority if specifically requested by the UN. The captain of the *Plymouth* therefore allowed the tanker to proceed and on April 5, 1966, the *Joanna V* dropped anchor just over a mile offshore from the port of Beria.

The *Joanna V* became a symbol of defiance and a trailblazer for a long stream of tankers. *The Sunday Times* of April 10, 1966, reported through one of its sources that AG Morrison had signed a contract for as many as twenty-seven cargoes of crude oil totaling 3.2 million barrels, which would have been adequate to fuel Rhodesia for a year.

The *Joanna V* was being closely followed by a second tanker. It too had sailed from the Iranian port of Bander Mashur on March 27, 1966, with a cargo of crude for Beria. Before its departure from the Iranian port, it had also been purchased by Vardinyannis's Panamanian company, Varnima Corporation, and, like his earlier acquisition, it sailed under the Greek flag. The vessel that had been named the *Charlton Venus* was renamed the *Manuela* after the owner's wife.

Two days after the *Joanna V* docked at Beria the *Manuela* sailed past Beria. Its captain stated that he was on his way to Rotterdam via Durban, but on April 9 the *Manuela* suddenly turned and headed south again toward Beria. These two tankers had created the most serious foreign crisis that Prime Minister Harold Wilson had faced. The *Joanna V* had arrived at Beria and might empty her holds, while the *Manuela* was still on course for the port.

The prime minister had to preempt African demands for the use of force by taking a tough stand over sanction busting. If not, the British government would soon be facing even greater pressure at the UN. On the morning of April 7, 1966, the cabinet met and decided to seek authorization from the UN to stop the delivery of oil to Beria. Lord Caradon, Britain's representative at the UN, was instructed to call an immediate meeting of the Security Council. At the UN Secretariat the Council's fifteen members were prepared to meet, but the president for the month, the Representative of Mali, was nowhere to be found, and under the UN Charter the Security Council could not formally meet without its president. Mali appeared to be delaying the UN meeting out of anger that Britain had originally refused to use force to quell the Rhodesian rebellion immediately after UDI.

This delay by Mali in convening the Security Council frustrated Britain in that it still had no authority to prevent the *Joanna V* from unloading or the *Manuela* from entering Beria. When the Security Council finally assembled on April 9, Lord Caradon opened the debate by stressing the urgency of the crisis and submitting a draft resolution that would authorize the United Kingdom to use military force to prevent the arrival of oil tankers at Beria.

The African members of the Security Council wanted tougher sanctions; they also questioned why Britain had not used force to crush the rebellion at the time of UDI. Furthermore, they asked why British efforts were concentrated only on the supply of oil through Beria. If the delivery of oil to Beria represented "a threat to the peace" it was argued, then surely this should also apply to oil supplied from Beit Bridge and Lourenco Marques. The African members of the Security Council therefore tabled a number of amendments to the British draft; none of the African amendments obtained sufficient votes for approval, and the original British draft resolution was passed by ten votes to none, with five abstentions.

The Beria resolution called for the Portuguese government not to allow oil for Rhodesia to be landed at Beria or to permit oil to be pumped through the pipeline to Umtali. All states were instructed to divert "any of their vessels reasonably believed to be carrying oil destined for Rhodesia." More importantly, the resolution called upon Britain

> to prevent by the use of force if necessary the arrival at Beria of vessels reasonably believed to be carrying oil destined for Rhodesia, and empowers the United Kingdom to arrest and detain the tanker known as the *Joanna V* upon her departure from Beria in the event her oil cargo is discharged there."
> (Security Council Resolution 221, 1966)

The Beria resolution marked a turning point in UN history. This resolution was the first occasion since the Korean war that on which a dispute had been declared a "threat to international peace," giving the Security Council, under Chapter VII of the UN Charter, the right to authorize the use of force. The resolution determined that supplying oil to Rhodesia would constitute a threat to the peace.

With the passing of the Beria resolution on April 9, the Royal Navy acted fast and on the morning of April 10, 1966, the frigate *Berwick* intercepted the *Manuela* in the Mozambique Channel and forced her to proceed on a southerly course toward Durban. The next day the *Joanna V* docked at Quay Number Eight. This quay was ten yards from the pipeline; it was believed that the tanker was attempting to discharge its cargo, but by this time the ship was an international pariah.

At Quay Number Eight the Greek consul informed the captain that the Greek authorities had withdrawn the tanker's registration five days earlier and that the *Joanna V* was now a pirate ship. A few hours later the word *Piraeus*—the old port of registry—was painted over by a crewmember and replaced with the word *Panama*. However, the Panamanian government also withdrew the tanker's provisional registration. The same painter again blacked out the word Panama. It is clear to see how easy it is for owners to switch flags so as to attain certain objectives; however, in this case the situation had become untenable. Nevertheless, the use of open registries is still used to evade national jurisdictions.

On April 16, 1966, Ian Smith announced that Rhodesia would not use the oil from the *Joanna V*. His decision, he explained, was in order to avoid drawing Portugal into the growing crisis between Britain and Rhodesia. He stated that his country had only begun using the pipeline from Beria in 1965. "Prior to that we got along very well using other means"—this was a reference to the importation of refined oil products through Lourenco Marques—"and so we will make do and continue by using these other traditional lines of supply."

The *Joanna V* left Beria on August 18, 1966, fully laden with 110,000 barrels of crude. However, the tanker was not able to leave port safely when fully laden because a sandbank blocked the entrance to the port. The British government therefore gave permission for 16,000 barrels to be pumped into another tanker. The *Joanna V* was then escorted by the Royal Navy to international waters. The Royal Navy nevertheless continued patrolling the Mozambique Channel. There were ten interceptions during the rest of 1966, and fourteen in the following year. Over time, the number of interceptions declined, and in 1972 no tankers were stopped. The number of patrolling vessels was reduced, and from March 1973 the patrol was only operated on an intermittent basis. Of course, by this time the risk of a tanker trying to make for Beria was small, since the Umtali refinery could only become functional after a long recommissioning period.

The patrol was formally ended on June 25, 1973 the day of Mozambique's independence, for the newly independent state had undertaken to ensure that crude oil supplies to Rhodesia were not resumed without authorization from the United Nations Security Council. This ten-year exercise, which had by 1975 stopped fifty-two tankers—none of which had been en route for Beria with crude—was an expensive endeavor. The total cost involved seventy-six ships and twenty-four thousand men are difficult to estimate, but they probably exceeded one hundred million pounds.

All in all the Beria patrol was pointless and wasteful, for the patrol had attempted to deter any importation of crude to Rhodesia, but it did absolutely nothing to halt the flow of refined oil products. Dr. David Owen, as the spokesman on defense for the opposition at Westminster, pointed out that

> it is extraordinary that it was thought necessary to achieve this limited objective, of preventing crude oil reaching Rhodesia, in such a flamboyant manner, when a similar result could have been achieved by covert action, such as blowing up the pipeline or sabotaging the pumps.
> (Dr. David Owen, *The Politics of Defense*, Cape, 1972)

Dr Owen's suggestion could have led to an international incident with serious diplomatic implications; nevertheless, one wonders why a small military operation was not conducted inside Rhodesia to ensure that the Umtali refinery never resumed operations.

While efforts to stop the *Joanna V* from delivering and discharging crude oil at Beria, massive supplies of refined oil products were being sent into Rhodesia by other routes. African members of the UN were incensed at why Britain was ignoring the flow of refined oil products from South Africa and Mozambique, in light of the UN calling on Britain to "prevent by the use of force if necessary the arrival at Beria of vessels reasonably believed to be carrying oil destined for Rhodesia." The resolution therefore applied to both crude and refined oil. Nigeria therefore proposed a Security Council resolution calling on the United Kingdom to "prevent any supplies, including oil and petroleum products from reaching southern Rhodesia" (Security Council debate, May 17, 1996). However, this proposal was defeated, having failed to win sufficient support because Britain and other Western members of the Security Council abstained.

An oil embargo that did not include measures of stopping refined oil from entering Rhodesia from South Africa or Lourenco Marques was an exercise in futility. By the time the Beria Patrol was in place most of Rhodesia's oil was being sent by rail along the South African loop route. Railway tank cars were loaded with refined oil at the port of Lourenco Marques, and then transported into South Africa. Once inside South Africa via the border at Komatipoort, the tank cars were suddenly turned back the way they had come and returned to Mozambique. Here they were consigned to Rhodesia as if they had originated from South Africa. This intricate route made it difficult to identify exactly who was involved in supplying Rhodesia's oil.

The usual suspects—the oil companies Shell and BP—were implicated. These companies would sell a consignment of oil to the South African firm of Parry Leon and Hayhoe, who would then arrange onward shipment to Rhodesia. Through this loop route, seven hundred thousand barrels of oil were dispatched from South Africa in 1966, and around two-thirds of this oil was actually sent by the South African subsidiaries of Shell and BP.

The South African loop was much cheaper than road transport, but it was still expensive because the greater mileage involved additional transport costs. Customs duties also had to be paid on the oil when it crossed into South Africa, and the shipments took longer to reach their destination. Genta, the Rhodesian procurement agency, therefore suggested that the oil should be railed directly from Lourenco Marques into Rhodesia, and in May 1966 Sonarep sent the first direct shipment of twelve railway cars. There was no reaction from the British government and, not surprisingly, other international oil companies followed Sonarep's lead.

It is important to say something about Genta—the name is an anagram of agent. Genta was set up in February 1966 to coordinate the importation of oil in light of the sanctions being introduced. Its role was to purchase oil through Freight Services Limited, a company that acted on behalf of one or more oil companies and which eventually absorbed Parry Leo and Hayhoe, then distribute oil products to the Rhodesian subsidiaries of international oil companies for local marketing.

Following Sonarep's lead, Total's South African subsidiary dispatched its first consignment of fifteen railway tank cars for Mozambique to Rhodesia later that May. Mobil later became the first American-owned company to use the direct route. The case of Mobil is interesting in that, although oil was railed directly from Lourenco Marques into Rhodesia, the company set up a complicated paper trail to disguise the incriminating links between the oil companies and Rhodesia. The paper trail was complex.

Mobil South Africa sold its oil to Freight Services (due to South Africa's declared neutrality, it could therefore trade with Rhodesia without fear of legal complications), which would then sell the same consignment of oil to Genta; finally, Genta would sell the shipment to Mobil Rhodesia for distribution to the public. Payment would go back along the same channel.

Caltex also supplied oil via the direct route. Caltex also faced an additional logistical problem, for its South African refinery was located at Cape Town, nearly thirteen hundred miles away from Lourenco Marques. A swap arrangement was therefore put in place between Shell/BP and Mobil in Durban, under which Shell/BP and Mobil refineries would supply oil products to Caltex in Lourenco Marques for transportation by rail to Rhodesia. In return, Caltex would provide the other companies with matching quantities of oil to market in the Cape Town area. This arrangement ensured everyone saved on transport costs.

Shell and BP had problems that were not shared by their American and French counterparts. First, Shell Mozambique was registered in Britain, and consequently subject to the UK sanctions order. Second, most of the directors of Shell Mozambique were British citizens and therefore under legal obligation to respect the embargo. On the other hand, the operations of Total, Mobil, and Caltex were conducted in Mozambique by their South African subsidiaries, which were registered in Cape Town. Regardless of these problems, Shell and BP followed the other oil companies, starting direct shipments on December 8, 1966. The delay by Shell and BP in joining its rivals in using the direct route came about because these companies already supplied around two-thirds of the oil sent via the South African loop route. However, the direct route had its advantages and it made good business sense to use it instead of the loop route.

During 1967 a total of 10,796 railway tank cars were dispatched to Rhodesia via the direct route. Interestingly, the market share each local subsidiary of an international oil company controlled (except for the inclusion of Sonarep) remained the same as at the time of UDI. This suggests cooperation and collusion between the oil multinationals to preserve their stake in Rhodesia. Shell and BP supplied forty-1 percent of the oil, Mobil 18 percent, Caltex 16 percent, Sonarep 13 percent, and Total 12 percent.

The South African subsidiaries of two other oil companies are believed to have supplied lubricants to Rhodesia. One is Esso South Africa. The other was Castrol; its products continued to be marketed in Rhodesia from the UDI onward. The question of whether Castrol or its parent, Burmah Oil, might have contravened the sanction order was a sensitive issue, because, at the time, Dennis Thatcher, husband of the British Prime Minister Margaret Thatcher (1979–1990), was a director of Castrol from 1967 to 1976.

On February 15, 1968, President Kaunda held a press conference at which he claimed that between January 1966 and August 1967 a total of 5.7 million barrels of oil destined for Rhodesia had been delivered at Lourenco Marques. Among the guilty tankers were BP's *British Flag,* the Caltex *Cardiff,* and the Mobil *Energy.* He accused Britain of being among the "gangster nations" that were breaking the Rhodesian oil embargo (*Guardian,* February 16, 1968).

The embargo appeared ineffective, but during 1973 it looked like the sanctions might bite. This was due to the Arab-Israeli war of that year. Many African countries backed the Arab struggle against Israel, and in return the Africans won greater Arab support against minority rule in southern Africa. On November 28, 1973, a summit conference of Arab nations imposed a complete embargo on Rhodesia, South Africa, and Portugal. This caused problems for Rhodesia, including the rationing of gasoline, but after a few hectic weeks from 1973 through to early 1974, the Rhodesian discovered that they were able to import all they needed through the help of South African subsidiaries of multinational oil companies.

The use of sanctions to foster a peaceful solution to UDI was a failure in Rhodesia and this became evident on December 11, 1978. On that day, three big explosions ripped through the Shell/BP storage tanks in Rhodesia. The resulting fire quickly spread to the Mobil, Caltex, and Total tanks. After the fire raged for six days, it was discovered that twenty-eight of the thirty-five storage tanks had been destroyed. Over half a million barrels of refined oil products worth some ten million pounds were destroyed.

This chapter has shown how the transportation of crude and refined products by international oil companies frustrated the UN embargo against Rhodesia. The definitive account of how these oil companies and the British government itself had defied the sanctions is beyond the scope of this book. For a comprehensive analysis on sanction busting I recommend reading *Oilgate - The Sanctions Scandal* by Martin Bailey. Nevertheless, it is important to highlight how important the sourcing and transportation of oil is to an economy.

6

MARITIME SECURITY

The previous chapter mentioned the role of the Royal Navy in setting up the Beria patrol. For the sailors, the Beria patrol was an unpopular assignment; for the sailors saw firsthand how unnecessary it was to patrol Beria while oil was being shipped in through Lourenco Marques. However, the Mozambique Channel has always been of strategic importance to a number of countries.

During the 1970s and 1980s, there was considerable concern over the security of the world's sea-lanes and their attendant chokepoints. Moscow's desire to acquire military facilities in Mozambique was interpreted in part as a desire to control the Mozambique Channel, a chokepoint on the critical East Africa/ Western Europe/U.S. mineral trading route. Therefore, although the Beria Patrol came under considerable attack in the British parliament, the patrol was kept up until the independence of Mozambique; withdrawing the Navy might have offered the Soviet Union an opportunity to move its naval units into the Mozambique Channel.

Since the end of the Cold War, concern for sea-lane and chokepoint security has fallen. Nevertheless, the security of global maritime trade remains as critical as ever. There are hundreds of chokepoints of regional and local economic importance, but fewer than two dozen of them are on the world's international maritime trade routes, endowing them with global economic significance.

Vital World Chokepoints

Eastern Pacific, Europe, Africa, the Americas, the Mediterranean, and the Arabian Gulf

Bosporus Strait of Malacca Great Belt Mozambique Panama Canal
Dardanelles Sunda Strait Kiel Canal Channel Cabot Strait
Suez Canal Lombok Strait Dover Strait Florida Straits
Strait of Hormuz Luzon Strait Strait of Gibraltar Yucatan Channel
Bab-el Mandab Singapore Strait Windward Passage
Makassar Strait Mona Passage

The transportation of oil by sea hinges on its regard of the sea-lanes and chokepoints as scarce natural resources. In other words, the quantity of sea routes available or accessible falls short of effective demand. Despite advances in maritime technology, prevailing winds, ocean currents, and predominant weather patterns still determine the safest and most efficient trade routes.

Certain parts of the oceans are off limits during certain times of the year due to the threat from waves of severe destructive force. Zones of violent wave activity exist in the Atlantic and North Pacific during the winter, primarily between latitudes 50°N and 60°N (including the British Isles and North Sea countries), and in the corresponding latitudes during the summer in the southern ocean (affecting the increasingly used Cape Horn and the Strait of Magellan routes). Similarly, ships transiting the Indian Ocean, the tropical southwest Pacific, the West Indies, and the China Sea during the monsoon season may also encounter waves of destructive force to damage or sink even a modern vessel.

Chokepoints can also be considered scarce resources because of the growing volume of global maritime trade. This has exerted pressure on existing sea-lanes and choke points because more nations than ever now belong to the global capitalist trading system, and most of the world's international trade moves by ship. However, as ship design and technology have allowed for larger, faster, stronger ships, some ship owners have become more willing to transit their fleets through increasingly dangerous waters. For example, the Cape of Good Hope (around southern Africa), the straits of Magellan and Cape Horn (around South America) and to a lesser extent the Northwest Passage across North America are now routinely used most months of the year.

This chapter is mainly about security of the sea-lanes, but it is also important to address the safety of the tanker itself, for at one time it was believed that the mere size and weight of a loaded tanker was the one thing afloat that was closest to being an imperturbable and relatively immovable object in violent weather. As with the Titanic, size was believed to hold its guarantees. However, what was overlooked was actual wave impact upon such massive floating objects. A loaded supertanker should be regarded as a heavy stationary object in the water rather than as a buoyant vessel able to ride the waves. It is important to understand the force of the waves.

Thomas Stevenson found in 1843 and 1844 that summer waves upon Skerryvore Rocks had a force of 611 pounds per square foot, and winter ones 2,086 pounds per square foot. The Skerryvore measure indicates the wave power that a low-lying tanker faces in bad weather. However, recently, a few tankers have been fitted with actual automatic devices able to measure the wave impact upon the bows and to warn the captains, oblivious in their towers a quarter of a mile back. When deeply laden, the waves fall upon tankers with the same impact they do upon rocks because the ship is not rising with them but bashing against them.

The Southern Seas

The sea-lanes of the southern seas are of great importance to Africa. I have therefore highlighted some of its salient features. The wild nature of the seas off southern Africa should not be underestimated, and they are a considerable threat to shipping. This is due to the current running down the east coast of Africa. The "Africa Pilot," the admiralty guide that every ship navigating round that continent's coasts should have aboard, provides a good hint of the serious trouble to expect. "In the event of meeting a southwesterly gale off this part of the coast, a very dangerous sea will be experienced at or outside the edge of the one hundred fathom line."

It is not unusual for a ship sailing eight to ten miles off the shore to suddenly find that it is falling into an apparent hole in the sea. The bow plunges into an abyss, and then at the foot of the trough anywhere from forty-five to nearly sixty feet high waves crash upon the vessel. These monstrous waves off the South African coast are created by a peculiar combination of seabed characteristics, the Mozambique-Agulhas Current, and weather. The East African continental shelf ends abruptly a short distance offshore, with a cliff-like edge above the great deeps.

The Mozambique-Agulhas Current passes along the edge of this undersea escarpment and, with its speed of four to five knots, creates surges similar to the effect of a gigantic flow of rapids. In winter the westerlies that blow from the far south whip around the corner of Good Hope and become southwesterly up the coast, creating their own huge seas after having accumulated their force over the long distance they have traveled before running into the surge of the Agulhas Current, generating massive waves.

The waves created by this powerful collision between current and gale winds bring about a spectacular phenomenon. The wave systems superimpose upon each other and thus frequently create a wave of extraordinary height and power at whose base lays a sudden, deep gulf. When a ship falls into this gulf, it is then followed by a massive wave upon its back; it is certainly this sort of wave that broke up the *World Glory* that spilled 46,000 tons of crude, the 70,000-ton *Wafra*, and the 100,000-ton *Texanita*.

These waves are at their worst along the one hundred fathom line, which is more or less the course of the edge of the offshore shelf, where the main flood of the current runs. This line varies from eleven to twenty-two miles offshore. Tankers traveling along that coast are separated into two lanes, as in other busy waters. This two-way traffic system was provisionally introduced by the Intergovernmental Maritime Consultative Organization (IMCO) early in 1971, and because it put loaded tankers on the inside lane, closest to the coast, this had inherent dangers in that any tanker in trouble will not only face the problem of the sea but also the perilous shore. Some experts feel that fully laden tankers should be running farther out at sea. However, ship owners prefer the inside lane, because the Agulhas Current represents a major economy for loaded tankers: assisted propulsion for two or three days at no cost. It is important to point out that the two shipping lanes are only two miles apart, and that the whole stretch of water is dangerous, although some parts are clearly much worse.

Following the closure of the Suez Canal after the Arab-Israeli war of 1967, by 1968 tankers were into their second southern winter, and accidents were occurring with disastrous effects on the maritime environment.

Oil tanker disasters started in February 1968, when the 81,000-ton French tanker *Sivella* ran aground off a Cape Town suburb, a few miles from the harbor. A few weeks later the 48,000-ton German tanker *Esso Essen* struck a submerged object and ripped its tanks, spilling more than 4,000 tons of crude into the sea off the coast of Cape Town. While the *Esso Essen*'s oil was polluting the coast, a 16,000-ton Greek tanker—the *Andron*—was foundering in a storm off the Cape. It too released large slicks of oil into the sea. As if this were not enough, that winter the fully laden *World Glory* broke up and went down north of Durban, sending its own sixty-mile long slick south to the Agulhas Current.

The damage done by these incidents (and they were by no means the only that season) continued in later years, causing considerable damage to the marine environment. The issue of pollution will later be discussed more fully. What is important here is to note that the security of the tanker translates into security or protection of the marine environment, chokepoints, and sea-lanes.

As I continue to concentrate on the sea-lanes around the Cape, I should note that in early 1969 a tanker oil spill killed the penguin population of Dyer Island, near Cape Agulhas. In November 1970 the 46,000-ton *Kazimah* spilled two hundred tons of oil sludge when it went aground in Table Bay. The resultant slick went mainly north on the Benguela Current, and a number of penguins were killed. The *Wafra* spill three months later killed between three and four thousand penguins. August 1974 was a particularly bad month along the coast of South Africa. After the grounding of the *Oriental Princess,* the 20,000-ton Norwegian tanker *Produce* struck a reef off Durban and sank. The survivors were picked up by the freighter *SA Oranjeland* that, after taking them to the port of East London, went aground. Three losses of large ships within three weeks was bad enough, but had the VLCC *World Princess*—which became disabled at the same time—gone aground as well, it would have represented one of the most disastrous sequences of shipwrecks on record on any coast.

Part of the reason behind these accidents has to do with overloading. Some analyses of tanker losses under severe conditions conclude that the losses occurred principally because the ships had been poorly loaded, or because cargo and ballast had been poorly distributed.

Plimsoll Line

The Plimsoll line or mark is a load-line marking on the side of a ship's hull. It shows the draft to which a ship can be safely loaded under certain conditions (e.g., winter or summer). A tanker "loaded down to the Plimsoll line" carries the maximum weight of cargo; any more cargo would lessen the vessel's chances of a safe voyage. Every ship has its Plimsoll line; it can be seen in the familiar red "boot topping" that covers the ship's bottom as far as her water-line. As a ship discharges, the red line rises until it stands high above the sea—especially in tankers, which lie so deeply in the water when loaded. The ship is then said to be riding high.

The line was largely the work of Samuel Plimsoll, born in 1824, who made improvement of conditions at sea his life's work. His main target was the nineteenth-century habit of sending unseaworthy and overloaded yet heavily insured ships to sea hoping that they would sink so that the owners could collect on them. Even when insurance fraud was not the aim, overloading was so common that many ships were lost because of it.

The idea of a load line is not new, the Hanseatic traders limited their loading, as did the Arabs in their dhows and the Venetians, who marked their load lines by the sign of the cross; they all realized that overloaded ships were not only slower but also dangerous.

Plimsoll was elected to parliament in 1868, where he was active in promoting the Merchant Shipping Act of 1890. This act made it compulsory for every British ship to have a permanent load line painted on the vessel's hull to indicate the depth to which it could be loaded. Load-line regulations for American ships are regulated by the United States Coast Guard as provided for under the Load line Act of 1929.

At an international conference in London from 1929 into 1930, Plimsoll's line and other safety measures were modified into a modern code known as the International Safety of Life at Sea and Load-Line Conventions, and these served (with periodic additional modifications) as the international guide for the safe management of ships at sea. Any alteration or suspension of these codes would appear irresponsible.

However, in March 1966 a new international load-line conference was held in London at which it was decided to amend the load lines for tankers passing the Cape during winter. Tankers already had permission to load deeper than usual, but these tankers still had to ensure that they were not so deeply loaded that their winter marks were below the water as they rounded the Cape in winter. This conference removed this provision.

The argument was that it was unreasonable to demand that a tanker should load less oil because for five days out of a thirty- to forty-day passage required this lighter load. Yet those five days and the sort of weather and sea conditions prevalent around the Cape are precisely what the rules were for in the first place.

These changes were based upon apparently valid and logical calculations. All deadweight calculations for tankers are based on the assumption that the temperature of a cargo of crude is 60°F. In the Arabian Gulf the temperature of oil being loaded can be much higher, thus the oil occupies more space than it would at 60 degrees. It is also important to allow for expansion in the trop-

ics. As mentioned earlier, crude oils vary considerable in weight; Middle Eastern crude tends to be much lighter than Chinese crude, for example. This
means that a tanker can sail from the Gulf lighter than its cubic capacity
allows.

Another point is that, as a tanker steams ahead, it is bound to get lighter;
therefore an overloaded vessel may start off overloaded and arrive at the destination port safely loaded. Anyway, at the time the rules were amended for the
Cape, VLCCs were still a novelty and experience with them was extremely
limited. The weights loaded into their holds were unprecedented; no one truly
knew (except theoretically) what their fullest strain might be upon those huge
hulls.

Anyhow, the request to amend the rules was swiftly accepted not only by
the governments of the principal maritime nations, but by insurers as well.
These amendments were made under the auspices of IMCO, which is now
called the International Maritime Organization (IMO). It is appropriate here
to give some background on the IMO or, as it was known during the amendments, the IMCO.

The International Maritime Organization

The International Maritime Organization, headquartered in London, is a specialized agency of the United Nations; it promotes cooperation among governments in matters involving international shipping. About 130 countries
belong to the IMO, and representatives from these countries attend IMO
meetings to deal with matters such as navigational safety, ship design and
equipment, crew standards, and pollution control. The agency was created by
a UN maritime conference in 1948 and began to operate in 1959 as the
IMCO; it changed its name to IMO in 1982.

What follows is an account on the role and workings of the then IMCO in
amending the Load-Line Convention. The IMCO/IMO is an important
international body—it is the only organization with some nominal jurisdiction
over maritime matters and thus, for form's sake at least, the only one that
could have sponsored the two amendments regarding the deeper loading of
tankers and the lifting of winter load limits at the Cape.

After a session of the IMCO's governing body, the assembly, a convention
is adopted that embodies its decisions. This is then to be ratified by domestic
legislation in the parliaments or legislatures of the IMCO members. A convention becomes international law when two-thirds of the IMCO's member-

ship has deposited articles of ratification. This usually takes years. However, a shortcut exists.

This is done by tallowing a diplomatic conference to stipulate in its final convention how many nations are required before the convention comes into force. A number is set, and when this group of nations has ratified the convention, it becomes law between them. This number is not fixed in the IMCO's charter, but is decided by every conference for every different convention. It must include at least seven nations with at least one million gross tons of shipping each.

It goes without saying that if any edict is thus recognized as the law of the sea among such a sizable group of maritime powers, then it is going to be accepted by others who find it convenient to do so. The Load-Line Convention is one that might be said to have suited everyone. Conversely, even when a particular law is in force among a few nations, those who are not too eager for its principles can defer applying them, if they so wish, until it actually becomes international law. Unfortunately, most pollution laws have a tough time coming into force for this reason (more on pollution laws later).

At the conference called by the IMCO in 1966 to amend the Load-Line Convention by suspending the winter load lines for tankers rounding the Cape, responsibility for this task was given to a technical committee of shipping experts. These experts did not know the handling characteristics and responses of big tankers in the sort of weather that was normally encountered off South Africa in the bad winter season at the time.

The final convention stipulated that the load-line amendments could come into force once fifteen specified nations (Panama, Liberia, Britain, the United States, Russia, Tunisia, Trinidad, France, South Africa, Malagasy Republic, Peru, Somalia, Denmark, Israel, and the Netherlands) had ratified it. It took the fifteen states a year to do this, and according to IMCO rules a convention comes into force twelve months after the last such ratification, which was received on July 21, 1967. The convention went into force in July 1968.

This convention dismayed the South African port authorities, and this is reflected by reports published on June 15, 1968, after the *World Glory* disaster and immediately before implementation of the IMCO amendment on load lines. The report was written by George Young, a respected authority on maritime affairs in general and on the southern seas in particular.

> One of the greatest paradoxes in the international organization for the safety of life at sea has been the authority granted tankers of major West-

ern powers rounding the Cape in recent months to overload their vessels. And after 21 July the long-established rule for higher loading of ships in winter in these waters is no longer applicable. After the closing of Suez the scarcity of oil in Europe induced controlling authorities to advise their shipmasters to ignore winter load-line rules off South Africa, and to come through these waters down to summer marks.

When associated with an earlier authority, which allows tankers to operate with less freeboard—representing another five thousand tons of cargo—the tankers have been negotiating winter storms in half-tide rock conditions Some have sustained structural damage, another had a spare propeller on the deck break loose and threaten the ship's safety, and this week the World Glory broke in two. Two weeks ago another Greek ship sank off Luderitz.

Conscientious surveyors in South Africa who have apprehended ships for overloading, have had to condone firm action when correspondence was produced from overseas authorizing ships to run in this condition.

Using the Cape as an example, it is clear that the safety of ships on this route is important not only for the vessels but also for the entire Southern ocean for much of the world's oxygen. By far the largest single portion of the life of the world's seas either exists or is dependant upon the southern ocean, whose fauna and flora are the richest and most prolific.

Moving on to the security of chokepoints, note that they are under stress due to increased demand. For example the Panama Canal, which is already operating at full capacity, is anticipated to handle an 18.5 percent rise in traffic by 2010, and a 48 percent rise from current levels by the middle of the century.

The U.S. Department of Energy anticipates that world demand for oil will grow by 44 percent between 1995 and 2015, with most of the increase being met by the Arabian Gulf nations. This raises security concerns that can take two forms: physical constraints that restrict passage, and actions by states or non-state actors that threaten or restrict free passage through the choke point. Today the predominant maritime power is the United States, one of the few countries that are able to physically provide security to chokepoints that directly affect its economy. A chart follows, showing how some chokepoints can be considered a scarce resource that needs U.S. protection because of increased demand.

Security Concerns Due to Increased Demand
Choke Point Importance to the U.S. Demand
Turkish Straits Black Sea & Caspian Sea Oil 1996: 123 ships/day
1997: 137 ships/day
Strait of Malacca Asian/Oil Trade 2000: 200+ ships/day
Sunda Strait Asian/Oil Trade: 3,500 ships/year
Lombok/Makassar Straits Asian/Oil Trade: 3,900 ships/year
Singapore Strait Asian/Oil Trade 1997: 200 ships/day
Strait of Hormuz Global Oil/Trade 1997: 60 ships/day

One of the main concerns regarding choke points is that many of them are extremely narrow, often only a mile or two at their narrowest point. Some are so narrow (the Turkish straits, for example) that they have to be closed to two-way traffic when the largest ships are in transit. In other cases, navigation difficulties of the larger ships through the narrowest straits are compounded by attendant tide, current, and wind extremes. Depth is also a paramount issue, particularly where silt begins to accumulate.

Depth is a serious issue for the Suez Canal. Fully laden VLCCs require transit depths of sixty-eight to seventy feet. Dredging the Suez will only achieve a depth of sixty-two feet. Depth is an issue for the Panama Canal too; currently the canal can only accommodate ships of up to 65,000 tons. This means big ships have to use the longer and more dangerous Cape Horn or Strait of Magellan routes. During the 1997–1998 El Nino event draft restrictions were particularly severe. In normal years, maximum allowable draft in the canal is 12.04 meters. During El Nino, which resulted in drought conditions in the canal zone, the maximum allowable draft had to be lowered to as little as 10.52 meters.

Accidents are also a major consideration in many chokepoints, as near misses or actual collisions or groundings involving large tankers occur. These worries are not without foundation; between 1988 and 1994, the number of collisions increased in the Turkish Straits, including a major oil spill in 1994 that burned for a week in the narrow straits. In 1996, Turkish Energy Minister Nusnu Doyan said that the amount of oil transited through the Bosporus could be transported safely only by allowing an additional increase of 20 percent; more than that could mean closing the straits to two-way sea traffic for eight hours "almost every day."

Another security concern comes about due to the behavior of a few nations that have threatened free passage through some of the world's chokepoints.

Those most affected include the straits in and bordering the South China Sea, the Strait of Hormuz in the Arabian Gulf, and the Turkish straits of the Bosporus and Dardanelles. In each case, littoral powers have threatened free passage. State challenges of free passage are of two forms: state military actions and disputed state claims.

State Military Actions

History shows that wars have been declared in the name of protecting maritime trade routes. For example, in 1956, when Egyptian President Gamel Nasser nationalized the Suez Canal, British and French troops along with Israel invaded the Canal Zone. This war lasted only a week, and invading forces were withdrawn within a month.

In 1967 the Suez was at the center of another conflict. Egypt had banned Israeli ships from the Suez Canal since May 1948, and in 1967 Egypt blockaded the Gulf of Aqaba, Israel's only access to the Red sea. Israel responded, resulting in the Six Day War. On June 6, 1967, Egypt closed the Suez Canal and broke relations with the United States; United Nations Resolution 242, passed in its aftermath, laid down the principles for Middle East peace; one of its conditions was free navigation for all ships through international waterways such as the Suez Canal. However, the canal remained closed for the next fifteen years due to continued military activity in the canal region.

During the Iran-Iraq war of the 1980s (more on this topic in next chapter), military actions between the warring parties threatened the passage of tankers through the Arabian Gulf and the Strait of Hormuz. Iran's threats to attack oil tankers were of great concern to Kuwait, which then sought protection from the Soviet Union. This was unacceptable to the United States, which undertook the re-flagging of eleven Kuwaiti oil tankers and provided a naval escort for their transit through the Strait of Hormuz.

Another example of military action to keep a vital sea-lane open occurred with U.S. military intervention in Panama, known by the United States as Operation Just Cause, in December 1989. The primary objective of Operation Just Cause was to restore Guillermo Endora to power and remove General Manuel Noriega; a secondary objective was to keep the canal open. Noriega had, on more than one occasion, threatened to close the canal by sinking several ships in it.

Areas of particular concern around the world are the Arabian Gulf, the Strait of Hormuz, and the South China Sea. The Strait of Hormuz lies at the entrance to the Arabian Gulf. At the narrowest, the strait consists of two one-mile-wide channels for inbound and outbound tanker traffic, as well as a two mile-wide buffer zone. The strait is critical because over fourteen million barrels of oil pass through it yearly. In early 1995 Iran deployed some 6,000 troops and heavy weapons on Abu Musa (also claimed by the United Arab Emirates) and other islands at the entrance to the strait. The secretary of defense for the United States at the time, William Perry, said that the deployment, which included anti-ship silkworm missiles, could "only be regarded as a potential threat to shipping in the area."

Although relations between Iran and the United States have warmed a little, passage through the Strait of Hormuz is an issue of concern several times a year when Iran holds its war games at the mouth of the gulf. In April 1996 naval maneuvers by the Iranian navy had the effect of intimidating oil tanker owners into holding their ships back. This in turn had the effect of raising spot oil prices. The future developments in Iraq will in part determine the security of the region.

Regarding the South China Sea, security concerns arise because of China's claim in 1992 to 95 percent of the South China Sea as its territorial waters. International law recognizes only a twelve-nautical-mile territorial sea plus a two-hundred-mile exclusive economic zone; the area claimed by China extends up to one thousand miles from the Chinese mainland and includes Japan and the Philippines within Beijing's security range. This area includes the Spratly, Paracel, and Senkaku island chains, which China claim as its own and are contested in varying degrees by Taiwan, Indonesia, the Philippines, Vietnam, Brunei, and Malaysia.

The disputes over the Spratly Islands are of the most concern in that they lie directly in the path of shipping lanes that converge on the Indonesian Straits. These shipping lanes bear the traffic that transports oil from the Middle East to Japan and the west coast of the United States. This traffic is approximately one-quarter of the world's total shipping trade.

On more than one occasion China has used its armed forces to reinforce its claims. In 1995 there was a military buildup and standoff with the Philippines over Mischief Reef in the Spratlys. China has displayed its military might through large military exercises, such as those that were conducted at the same time as Taiwan's presidential elections in March 1996. In that case Beijing engaged in live-fire war games off the southeast coast near the Taiwan Strait involving more than ten warships and as many aircraft. Another example of Chinese force occurred in March 1988, when Vietnamese and Chinese forces clashed, resulting in the loss of two Vietnamese ships and the occupation of six islands by China.

Regarding Indonesia (the world's largest natural gas exporter), China has laid "historical claims" to the Natunas Islands, which are rich in oil and natural gas. In September 1996 the Indonesian armed forces conducted its most extensive war games in four years on the islands. This was a veiled message to China stating that the Natunas belong to Indonesia. Singapore is concerned about the South China Sea as well, and in July 1997 increased the number of its armed vessels patrolling the area with the intent of securing the sea-lanes.

In these incidents it is important to note that merchant shipping has been the direct target of state action. The main worry, however, is that, if conflict breaks out, ships will have to be detoured around the conflict zone. This loss, even temporarily, of some of the world's most important shipping lanes would disrupt trade, extend transit times, and result in higher oil prices.

Post-September 11, 2001

Global transport infrastructure, both as a potential target for terrorism and even more threateningly as a potential weapon of mass destruction became a reality on September 11, 2001. The link between transport infrastructure and terrorism was further enforced with the attack on the oil tanker *Limburg* of Yemen in October 2002 and the Madrid train bombings in March 2004.

Following the September 11 attacks the International Maritime Organization, as the United Nations agency responsible for the safety of international shipping, launched a concerted and thorough response to the possibilities of terrorism being directed against ships or terrorists seeking to use ships as weapons or using proceeds of shipping activities to fund further operations.

The IMO adopted a comprehensive near-regulatory regime at the end of 2002 which outlined in considerable detail what ship operators, ship crews, port authorities, governments, and other stakeholders involved in international shipping should do in order to prevent and minimize the threat of terrorism.

Unlawful acts targeting international shipping is not new; since the late 1970s, IMO has focused attention on attacks to shipping, including barratry, maritime fraud, and the unlawful seizure of ships and their cargoes. Since 1982 it has been monitoring acts of piracy and armed robbery against ships and has put in place modalities to combat unlawful acts in areas that suffer most from them. At the time of this writing the IMO has received some 3,500 reports involving loss of ships and, in many cases, loss of life. What is most disconcerting is that the level of violence appears to be on the rise.

In 1985 the Italian cruise ship *Achille Lauro* was hijacked by terrorists who killed a passenger before agreeing on terms to end their siege. This incident spurred the international community to come together to combat terrorism at sea. That same year, IMO's fourteenth assembly adopted a resolution on measures to prevent unlawful acts at sea that threaten the safety of ships, their crews, and passengers.

The Assembly invited IMO's Maritime Safety Committee (MSC) to develop detailed and practiced technical measures to protect passengers and crews on ships, taking into account the work of the International Civil Aviation Organization in the development of standards and recommended practices for aircraft security. The *Achille Lauro* hijacking eventually led to a circular that gave guidelines on the steps to be taken in reference to passenger ships engaged in international voyages and port facilities that service them.

Then, in November 1986, work began on the preparation of a convention that was more relevant to oil tankers and covered the subject of unlawful acts against the safety of maritime navigation. In March 1988 a conference was held in Rome that adopted the Convention for the Suppression of Unlawful Acts against the Safety of Maritime Navigation.

This so-called SUA Convention and its protocol relating to offshore platforms was

> to provide for the comprehensive suppression of unlawful acts committed against the safety of maritime navigation which endanger innocent human lives, jeopardize the safety of persons and property, seriously affect the operation of maritime services and thus are of grave concern to the international community as a whole.

The main purpose of the SUA is to put mechanisms in place against persons committing unlawful acts against ships, such as the seizure of ships by force. The convention obliges contracting governments to either extradite or prosecute alleged offenders. Since its adoption, the SUA Convention has gathered considerable acceptance and received sufficient ratifications to enter into force in 1992.

Since September 11, 2001, security considerations have led to a dramatic increase in the number of parties to the SUA Convention. In October 2001 the SUA convention had been ratified by 56 states and the 1988 SUA Protocol by 51 states. By July 2004 the convention had been ratified by 107 states, which between them were responsible for 81.52 percent of the world's merchant shipping tonnage, and the protocol by 96 countries, which between them were responsible for 77.66 percent of the world tonnage.

It is clear that after September 11, 2001 comprehensive processes were put in place to review all existing measures adopted by IMO to combat acts of violence and crime at sea. These initiatives to combat terrorism led to the Diplomatic Conference on Maritime Security held at IMO's London Headquarters in December 2002. The meeting was attended by 108 contracting governments to the 1974 Safety of Life at Sea Convention (SOLAS), and observers from two other IMO member states and the two IMO associate members. United Nations specialized agencies, intergovernmental organizations, and nongovernmental international organizations also sent observers.

The outcome of the conference was a new, comprehensive security regime for international shipping (which entered into force on July 1, 2004). A number of amendments to the 1974 Safety of Life at Sea convention were adopted, the most far-reaching of which was a new chapter (XI-2, on special measures to enhance maritime security), which enshrines the new International Ship and Port Facility Security Code (ISPS Code).

The ISPS Code applies to passenger ships and cargo ships of 500 gross tonnages and upward, including high-speed craft, mobile offshore drilling units. It also applies to port facilities serving such ships engaged on international voyages, detailed security-related requirements for governments, port authorities, and shipping companies in a mandatory section (Part A), together with a series of guidelines about how to meet these requirements in a second, nonmandatory section (Part B).

The diplomatic conference was referred to in the United Nations General Assembly, which adopted a resolution on oceans and the law of the sea that welcomed initiatives at IMO to counter the threat to maritime security from terrorism. States were encouraged to fully support this endeavor.

However, the mere existence of a new regulatory maritime security regime is no guarantee that acts of terrorism against shipping may be prevented and suppressed. It was also recognized that not all the member states of IMO have the same ability to implement the new measures, particularly in developing countries where there may be a shortage of expertise, manpower, and resources.

In this regard, the conference addressed the issue of technical cooperation and assistance by encouraging contracting governments to the convention and member states of the organization to provide, in cooperation with the IMO, assistance to those states that have difficulty in meeting the requirements.

Furthermore, the Integrated Technical Cooperation Program of IMO was called upon to further strengthen the assistance provided to developing countries. There is a need to address the future needs of developing countries in education, training, and improvement in port security measures. Donors, international organizations, and the shipping and port industry were urged to contribute financial, human, or in-kind resources to the Integrated Technical Cooperation Program of IMO for its maritime and port security activities.

As mentioned earlier, it is also important to keep strategically important shipping lanes secure and open to individual maritime traffic. One of the most important of these, not only to the tanker trade but also to other shipping, is the Strait of Malacca. It is worth reemphasizing that this 800 km long, narrow link between the Indian Ocean and the South China Sea is of paramount importance to global trade.

Tankers and bulk carriers transport oil, coal, iron ore, and grain (as well as other goods) to the manufacturing centers of Southeast and Northeast Asia, while high-value manufactured goods are transported in container ships back through the same outlet to feed consumer markets all over the world. Each year some fifty thousand ship movements carrying as much as one-quarter of the world's commerce and half of the world's oil pass through the Strait of Malacca and Singapore Strait each year.

That is why the security of this shipping channel is an important issue. Being an area of considerable activity, it is also a target of pirates and armed robbers. Unfortunately, Southeast Asia is still recording the highest number of pirate attack globally, and there is clearly a possibility that terrorists could resort to pirate-style tactics or even work in concert with pirates.

In this regard, the IMO has designed the Maritime Electronic Highway project for the Strait of Malacca and through the cooperation of the littoral states of the Strait of Malacca and Singapore Straits aim to ensure that this strategic lane remains open.

Another strategic lane is the Strait of Hormuz, and according to various estimates an average of three thousand ships sail through the strait. However, due to the Iran-Iraq war, attacks on international shipping from 1981 to 1987 led to what is now termed the tanker war.

7

TANKER WARS

Oil and its transportation has been one of the fundamental reasons why wars have been waged since the World War I. Japan's decision to go war and the involvement of the United States in World War II can be traced to the access of oil. Japan's attack on Pearl Harbor was part of a wider strategy, as Japan was also bombing the Philippines, Singapore, Hong Kong, taking over Thailand, and preparing to invade the East Indies all at the same time.

The attack against Pearl Harbor was mounted to provide protection for Japanese troops to invade the East Indies and the rest of Southeast Asia by destroying the American fleet and, thereafter, to protect the sea-lanes, most importantly the tanker routes from Sumatra and Borneo to the home islands. The main goal of this military campaign was the oil fields of the East Indies.

However, in executing this military campaign the Japanese overlooked the oil supplies stored on the island of Oahu. This island was not attacked, and yet almost five million barrels of oil were sitting there in surface tanks at the time Pearl Harbor was attacked. If the Japanese had destroyed this oil depot World War II may have continued for another two years at least.

Before the outbreak of war, the Japanese had already planned on extracting sufficient oil from the East Indies to mount a formidable fight against the Americans and British. This was a tricky strategy, for it depended on the integrity of Japan's own shipping system. Japan had entered the war with oil reserves sufficient to last two years, or so it thought. Beyond this time frame, Japan would get oil from the East Indies. This was to become Japan's Achilles heel.

Japan's Achilles heel was the vulnerability of Japanese shipping to submarines. Japanese military planners underestimated the abilities of American submarines. These subs were formidable weapons that broke the critical shipping links between the East Indies and Japan. By 1943 the Japanese began serious attempts to protect shipping against submarines, including the establishment of convoys.

Of Japan's total wartime steel merchant shipping, approximately 86 percent was sunk and another nine percent seriously damaged. Comprising less than 2 percent of American naval personnel, the submariners were responsible for 55 percent of the total loss, while other Allied submarines contributed another 5 percent.

Oil tankers were among the prime targets of submarines, and tanker sinkings rose very sharply from 1943. By 1944 tanker sinkings were ahead of new tanker construction. Consequently, the growing shortage of oil increasingly affected Japanese military capabilities. In order to stretch oil supplies, many Japanese ships burned unrefined Borneo crude, which was highly flammable and thus a threat to the ships it powered.

The Japanese could also do little to stop the flow of oil to the American forces in the Pacific. The Americans developed huge floating bases made up of fuel barges, repair ships, floating docks, and salvage ships. Roving fuelling task forces comprising two or three giant tankers plus destroyer escorts took up stations at designated areas where U.S. ships could refuel.

The British also had an overarching concern: how to secure oil for the war. War means much greater consumption of oil. The only place the British could get oil was from the United States. This was facilitated through the lend-lease that was instituted in March 1941; this meant that among those things to be lent for repayment at some undetermined time in the future was American oil. In the United States the neutrality legislation gradually loosened and lifted the restriction to ship supplies to Britain, but this supply route was vulnerable to attack.

These attacks came from the U-boats, and their favorite targets were oil tankers. Altogether, the number of tankers sunk in the first three months of 1942 came to almost four times the number built. In March 1943, U-boats sank 108 ships. However, in the last days of March things changed. The Allies broke the U-boat codes and successfully closed off their own convoy ciphers to the Germans. The tables had turned, and by May 1943 30 percent of the U-boats were lost at sea.

It is clear that oil and, consequently, tankers were major considerations in World War II in regard to Japan, the United States, and the United Kingdom. Germany's decision to go to war with the Soviet Union can also be traced to oil. From the very beginning, the capture of Baku and other Caucasian oil fields was central to Hitler's Russian campaign. During World War I a major German objective was the Ploesti oil fields of Romania, Europe's largest source of petroleum production outside the Soviet Union. In 1940 Hitler feared a strong Soviet Union would threaten the Ploesti oil fields.

This chapter's title is the name given to the Iran-Iraq War of 1980–1988 because of the deliberate targeting of tankers by the protagonists. However, as outlined earlier, targeting tankers and supply ships is not a new phenomenon. During the Iran-Iraq War, the significance of sea operations and the sustained campaigns waged by both protagonists against merchant shipping cannot be ignored.

During the war hundreds of merchant ships were attacked, more than four hundred seamen were killed, and millions of dollars worth of damage was suffered by insurers, charterers, and ship owners. The Tanker War showed how oil and the control of its export was a central feature that determined strategies and operations of both the Iranian and Iraqi armed forces.

The Tanker War is important in that its effects stretched well beyond the region itself, as vessels with foreign flags inevitably became unfortunate targets. It is odd to note that for six years, the international community did little to deter attacks on merchant shipping. Many analysts of the war feel that political, economic, and even military pressure should have been brought to bear on Iran and Iraq to halt anti-shipping attacks and ensure the free and safe passage of merchant shipping in the gulf.

This war highlighted a basic principle under international law, namely the right of innocent passage for neutral merchant shipping in both peace and war. This right was established a considerable time ago. The first universal treaty that established the right of neutral ships was the 1856 Declaration of Paris, which stated, "Neutral goods with the exception of contraband of war, are not liable to capture under enemies' flag."

Furthermore the Hague Convention of 1907 states "Belligerent[s] are bound to respect the sovereign rights of neutral powers and to abstain in neutral territory or neutral waters, from any act which would, if knowingly permitted by any power, constitute a violation of neutrality." The Tanker War ignored such international conventions with disastrous economic effects.

One effect was on the international shipping insurance market, by destroying the Hull War Risks Rating Committee and causing hull and cargo prices to rise and fall erratically. Most Lloyd's syndicates were not able to make profits out of war risk insurance, and marine insurers' total costs for the war ended up well over one billion dollars. Both protagonists were well aware of insurance rate vulnerability, and their attacks on international shipping were aimed not only at the ships but also at affecting rates charged in London and elsewhere, and thus the expense and profitability of continual voyages.

The root causes of the Tanker War pre-date the emergence of the modern states of Iraq and Iran and the discovery of oil. However, the discovery of oil raised the geopolitical and economic stakes in the region. Although Iraq is blessed with considerable oil reserves, the country has an extremely small coastline. This had economic and military implications, as it left Iraq extremely vulnerable to blockade. This is further exacerbated by the proximity of vital oil installations to the Iranian border. The only way to increase Iraq's coastline was by acquiring land from Kuwait or Iran. It was an open secret that Iraq harbored such ambitions.

Over the years since the fall of the Ottoman Empire a number of treaties were signed regulating the control of all the Shatt al Arab. However, the rulers of the empire were always careful to ensure that Iran did not lose its position in the area. This was further reinforced from the late 1960s, as the British and then the Americans supported Iran so as to ensure pro-Western stability in the region.

Iraq was also dealing with the Kurdish revolt that was being supported by Iran. To terminate this support, the Iraqis were forced to make concessions in the Shatt in the 1975 Algiers Accord, which was concluded by Iraq's then foreign minister and vice president, Saddam Hussein, and the Shah on March 6, 1975. This agreement shattered Iraqi ambitions of expanding its coastline. Instead of expanding Iraq's coastline, Iraqi access to the Shatt waterway had been constrained.

At the time Iraq invaded Iran in September 1980, Saddam's objectives included the complete capture of the Shatt al Arab and some oil-producing facilities. On the eve of the war, Saddam demanded that all ships in the Shatt must carry Iraqi pilots and carry the Iraqi flag. He did not rescind this demand for control of the waterway until August 1990, during the Kuwait crisis.

At the start of hostilities approximately seventy merchant ships were in the Shatt estuary, and a further eight were located in the Iraqi port of Umm Qasr. The first concern was for crew safety and how to extricate the ships. Within a few days reports of casualties grew. In the Shatt al Arab twenty-seven ships were damaged, fourteen seriously; of the vessels at Umm Qasr, three were damaged, two of them seriously.

Governments whose nationals owned the trapped vessels applied pressure for the United Nations to intervene. The United Nations initially sought Iraqi agreement to let the trapped ships leave flying the United Nations flag. Iraq rejected this proposal because it would undermine Iraq's perceived sovereignty over the waterways, bearing in mind that before hostilities broke out Iraq wanted all ships entering the estuary to fly the Iraqi flag.

United Nations Secretary General Kurt Waldheim requested Iraqi help in freeing the trapped ships on humanitarian grounds and without prejudice to the claims, positions, and rights of the parties to the conflict. Saddam refused, and President Bani Sadr rejected Kurt Waldheim's request for a local cease-fire to allow ships in the Shatt to leave under the United Nations flag.

In November 1980 the new United Nations Peace Envoy, Olaf Palme, engaged both parties to free the ships. Both countries did not want to alienate Palme, but the stakes were too high and neither country was ready to accede to his requests for a cessation of hostilities. Negotiations continued through 1981 with no positive results. Palme tried again in 1982, but it was now clearly unlikely that the ships would be freed.

With this realization, insurance brokers were forced to pay the owners of the trapped vessels. Special blocking and trapping clauses meant that if the ships remained trapped for over twelve months, then, for insurance purposes, they would be regarded as total constructive losses. About $450 million was ultimately paid by marine insurers for the trapped ships.

Toward the end of 1980 Iraq expressed further war objectives. Iraq indicated its desire to neutralize Iranian power in gulf waters by calling for freedom of navigation both throughout the gulf and in the Strait of Hormuz region. The international community grew concerned that Iran might react by closing the Strait of Hormuz with drastic consequences. A U.S. congressional research service report warned that closure of the strait would push the oil price to more than $100 per barrel, causing a "great price trauma for the world economy."

Closing the Strait of Hormuz would affect the interests of the Western world. Through the United States, these concerns were expressed by U.S. President Jimmy Carter on September 24, 1980. He stated, "Freedom of navigation in the Persian Gulf is of primary importance to the whole international community." He went on to say, "It is imperative that there be no infringement of that freedom of passage of ships to and from the Persian Gulf Region." The U.S. position was reiterated and emphasized in a statement to the United Nation Security Council on September 28, 1980, by stating "the freedom of navigation to and from the Persian Gulf, which is of primary importance to the international community, must not be infringed upon in any way."

To support this policy, Jimmy Carter increased the U.S. naval presence in the region, joining the European Community, so that by October there were approximately sixty American, British, and French warships either in the gulf of in the Indian Ocean. It is interesting to note that the U.S. naval presence was also clearly a move to counter Soviet power projection in the region.

Nevertheless, from 1981 to 1983 Iraqi attacks on merchant ships were sporadic and formed only a minor part of their broader campaign to disrupt Iranian oil production and export. During this time fifteen bulk carriers, twenty-five general cargo ships, and six tankers less than two hundred thousand tonnes were attacked. No tanker over two hundred thousand tonnes was hit, although an important observation is that after 1983 a shift began to take place, wherein tanker attacks increased.

During 1984 sixty-eight ships were hit: forty-nine by Iraqis and nineteen by Iranians, while four ships were damaged in collisions due to Iraqi attacks. Half of these were tankers/liquid gas carriers, indicating that tankers were now being specifically targeted by both countries and that the interdiction of oil exports had by 1984 become an explicit objective.

According to figures compiled by Lloyd's List, of an estimated sixty-seven ships hit, forty-eight were tankers under one hundred thousand tonnes, 39 percent were over two hundred thousand, and 12 percent of the tankers hit were more than three hundred thousand tonnes.

In 1985 the war was characterized as a battle of attrition, with Iraq escalating its offensives against oil installations and shipping. Iran replied to Iraqi strikes by its own attacks upon Iraqi towns and cities and increasing attacks upon shipping, particularly on Kuwaiti trade. By the second half of 1986 the Iranians were focusing intensely on Kuwaiti vessels. During this period, eight tankers owned by, or trading with Kuwait were attacked by Iran, leaving the Kuwaitis increasingly beleaguered and passing on their concerns to their gulf allies. The Kuwaiti foreign minister, Sheikh Sabah Al Ahmed Al-Sabah, expressed his dismay: "The Strait of Hormuz is not owned by Iran. It belongs to the whole world. It is not in the world's interest to interfere with shipping in international waters."

The foregoing notwithstanding, Iran's response was to stop and search merchant ships, especially those bound for Kuwait. The British government in the House of Commons stated: "Iran, actively engaged in an armed conflict, is entitled in exercise of its inherent right of self-defense to stop and search a foreign merchant ship on the high seas if there are reasonable grounds for suspecting that the ship is taking arms to the other side." By the end of 1986 the Iranians claimed to have searched more than one thousand merchant ships.

Even the United States recognized Iran's right under international law to stop and search ships. On January 12, 1986, the first American ship—the cargo ship *President Taylor*—was stopped, searched, and then allowed to proceed to Al Fujayrah with its cargo of cotton. This incident raised concerns in Washington because of the "danger of misunderstandings, overstepping of rights and norms, and even violence [that] are inherent in all ship search incidents." For this reason the U.S. Navy started to closely escort American merchant vessels in the gulf.

Kuwait found itself in an uncomfortable position, and on December 10, 1986, Kuwait made overtures to the U.S. Coast Guard concerning the possibility of registering merchant vessels under the U.S. flag. This request came soon after the failure of the U.S.-Saudi arms deal and the Iran-Contra scandal. It was hoped that Washington would feel compelled to intervene positively in the gulf to bolster its tarnished image in the world. Nevertheless, the U.S. response to the Kuwaiti request proved to be a turning point in the Tanker War.

In the context of escalating attacks upon ships flying the Kuwaiti flag, the Kuwaiti government began seeking foreign power protection by attempting to re-flag in China, France, the Soviet Union, and the United Kingdom. The Kuwaitis were particularly interested in registering seven tankers and four liquid petroleum gas (LPG) carriers (half the Kuwaiti tanker/LPG carrier fleet) in the United States. In December 1986 Washington was approached with this idea, but Washington was noncommittal. Decision-makers in the United States were wary of involvement in the Gulf. Although Gulf oil was important to the world economy, the reality of the matter was that tanker traffic to Kuwait was minimal and not worth the risk of a skirmish with Iran.

Kuwait, recognizing U.S. geopolitical policy in the region, approached the Soviet Union, probably as a way of galvanizing the United States to play a more direct role in the Gulf. On March 2, 1987, it was announced that three Soviet tankers, the *Makhachkala*, the *Marshal Bagramyan*, and the *Marshal Chuykov* would be leased to Kuwait. Then, on March 7, Washington finally agreed not only to register the Kuwaiti ships but also to escort them in Gulf waters as well.

A re-flagging agreement was concluded on April 2, 1987, but was formally announced on May 19, 1987. In this announcement, President Reagan stated, "The use of the sea-lanes will not be dictated by the Iranians. These lanes will not be allowed to come under the control of the Soviet Union. The Persian Gulf will remain open to navigation by the nations of the world." In order to comply with U.S. legal requirements, eleven of Kuwait's twenty-two tankers had their ownership transferred to the U.S. Chesapeake Shipping Incorporated; this allowed them to be protected by U.S. naval vessels.

Cold War rivalry with the Soviet Union, rather than the Gulf's economic importance, was the main reason for the re-flagging agreement. A convoy system would be used in escorting the ships after re-flagging; the rationale was that this would deter Iranian attacks, counter Soviet power projection, protect sea-lanes, and avoid attack involvement in the war.

London supported re-flagging, and in 1986 nine ships had been re-flagged on the British Register (none of them tankers), but in 1987 the number rose to eighty-five (including twenty-seven tankers) with another twenty-nine added by May 3, 1988. It must be stated that the international shipping community was not pleased with the re-flagging exercise. The International Association of Independent Tanker Owners (Intertanko) for example, supported a United Nations led protection scheme. The main concern was that re-flagging Kuwaiti ships would endanger ships that were not flying either the Union Jack or the Stars and Stripes.

Indeed, with naval powers such as the United States, the United Kingdom, France, and even the Soviet Union providing escorts, the Iranians adopted a policy of targeting ships from countries that did not, or could not provide naval protection. Countries such as Denmark, Greece, Norway, and Spain were among the victims. Before December 1987 Danish and Norwegian flagged ships were not attacked, but in the first half of 1988 suffered six attacks. In the next seven months there were eight attacks on Norwegian ships and Spanish ships were attacked on four occasions.

It is therefore not surprising that in March 1988 about a dozen ships switched to British registration in order to gain protection; most of these ships belonged to the U.S. oil majors Exxon, Mobil, and Chevron, which transferred nine ships to the British Register. It is important to note that no Kuwaiti ships were re-flagged in the United Kingdom, although some were re-flagged in Bermuda. This was for logistical reasons, as the United Kingdom's Armilla Patrol was overstretched and could not provide comprehensive naval protection.

On July 3, 1988, the *USS Vincennes* shot down an Iran Air Airbus A300 B2 airliner that was late on a scheduled flight to Dubai and for unknown reasons ignored all preliminary challenges by the *USS Vincennes*—more than 290 men, women, and children died. This incident had a great psychological effect on the Iranian government and, although bitter about the loss of innocent life, conciliatory remarks started coming from Tehran. Iran now realized that its people were fed up with war, a military victory appeared very remote, and the attacks on international shipping had helped isolate Iran in the eyes of international public opinion. Lastly, the war was adding to Iran's economic problems.

A major contributing factor leading to the Iranians to end hostilities was the decline in oil reserves for which some credit goes to Iraq's anti-merchant shipping campaign. In the first half of 1988 Iran earned $3.7 billion, nearly 25 percent less than the previous year. In contrast, Iraq's oil revenue had risen 25 percent. Iraq had the advantage of oil pipelines running through Turkey and Saudi Arabia, not to mention two closed lines to Syria. There was little that the Iranians could do to neutralize these exports. Interestingly, we have not heard the last of these pipelines, as they are likely to be a significant feature in the Iraq Oil-for-Food Investigation.

A cease-fire was accepted on August 8, 1988, but Iran said it would continue to inspect ships for Iraqi war materiel, and Iraq demanded freedom of navigation. This state of affairs threatened to breach the fragile peace. In response, Iraq dispatched the product carrier *Ain Zala* from a temporary offshore loading point to the southern Yemeni port of Aden with 18,000 tonnes of oil. The tanker reached its destination, and in October Iran announced the end of searches of neutral shipping.

In summarizing the Iran-Iraq War, or more appropriately the Tanker War part of this larger war, one notes that 411 ships were attacked by land, sea, air, and mine within the gulf and the Gulf of Oman. Many tankers were hit several times, and of the ships attacked, the vast majority (239, or 58 percent) were tankers, another 10 were liquid gas carriers. Together they made up 60.5 percent of the Tanker War victims. In terms of human costs, the Tanker War claimed 432 lives.

Ironically, the tanker market benefited from the war by the physical elimination of 10,483,840 in tanker tonnage from the market, which was a substantial surplus. From 1984 onward this stimulated demand in the secondhand market that made it profitable to maintain fleets even after the war. However, at the end of the war Worldscale rates were still not high enough to lead to new construction orders, but were high enough to allow ship owners to keep working older tonnage.

Two political considerations have come out of the Tanker War that are of great concern to the international shipping community. First, Saddam Hussein was able to attack merchant ships to bring international pressure upon the Iranian war effort and was vindicated and rewarded in using this strategy. Second, it appeared that the fate of sailors and the freedom of navigation were of subsidiary importance in the geopolitical strategic calculations of the leadership of the superpowers—despite public declarations. It took several years of attacks before the United States, supported by the Europeans, became fully involved.

Since the collapse of the Soviet Union and the events of September 11, 2001, in New York and Washington, the fight against terrorism has dominated the foreign policy engagements of the world's only superpower and its close allies (coalition of the willing). Ironically, terrorist attacks have increased, particularly in Iraq. Terrorism on transport infrastructure is a reality what concerns merchant shippers sailing in zones of conflict, be they in the South China Sea, the eastern Mediterranean, or elsewhere. Although in the Iran—Iraq War the United States and the Europeans were reluctant to intervene, the world since—September 11, 2001, warrants a more rapid and physical response. These would have to be worked out through the new comprehensive security regime for international shipping mentioned in the previous chapter.

8

OIL POLLUTION

Marine pollution has become a serious concern for the shipping industry, even though ships only account for a small proportion of damage to the marine environment. International scrutiny plus interest to oil pollution can be said to have started after the grounding of the *Torrey Canyon* in March 1967 off the southwest coast of England. This disaster, which at the time was the largest single boil spill in maritime history, was extensively covered through the international media.

The *Torrey Canyon* grounding led to eighty thousand tonnes of crude oil being spread along the British and French coasts, causing pollution over some two hundred miles. The environmental clean up cost claims brought in the United Kingdom amounted to some six million pounds and in France to some forty million francs.

The public concern that followed this disaster stirred governments and international organizations to come up with an antipollution agenda to counteract and address the environmental effects of major oil spills on the marine environment. It was the *Torrey Canyon* disaster that initiated discussions at IMCO (now IMO) that resulted in the introduction of two new international conventions.

The first came in 1969 with the 1969 IMCO International Convention on Civil Liability for Oil Pollution Damage, 1969 (Commonly known today as the CLC). This convention changed the traditional damage liability base from one of proven fault or negligence to that of a strict liability situation.

This convention doubled the ship owners' liability limits from those set out in the Convention on Limitation of Liability of Owners of Seagoing Vessels of 1957. Furthermore, it also introduced a system of certification, which had the effect of making oil pollution insurance compulsory and gave a direct right of action against the insurer of the ship owner who did not pay any compensation claims. The civil liberty convention entered into force internationally on June 19, 1975.

Another significant outcome that followed the establishment of the Civil Liability Convention was the adoption in 1971 through IMO of the International Convention on the Establishment of an International Fund for Compensation for Oil Pollution Damage (known as the Fund Convention). The Fund Convention established an international intergovernmental organization, namely the International Oil Pollution Compensation Fund (IOPC Fund), to manage the system of compensation.

The IOPC Fund, which entered into force in 1978, is the only worldwide intergovernmental organization that pays compensation to victims who have suffered pollution damage. It also relieves the ship owners of part of the financial burden imposed on them by the CLC.

The CLC applies to seagoing vessels of any type carrying oil in bulk as cargo. This means that dry cargo and any other type of vessel in ballast are not covered by the CLC. However, the convention will apply to a bunker spill from a tanker that is carrying oil as cargo.

The CLC applies in the event of pollution by persistent oil. The term *persistent oil* is not actually defined, but this includes crude oil, fuel oil, heavy diesel oil, lubrication oil, and even whale oil. Dirt ballast water discharge or any other polluting substances, such as chemicals, are not covered.

Another main feature of the CLC is that the registered owner of the vessel is the only one that can be held liable for oil pollution damage and no claim outside the CLC can be made against the owner or, indeed, the owner's servants or agents. Contracting states that are party to the CLC are required to ensure that all ships laden with more than two thousand tons of persistent oil in bulk as cargo that enter or leave any part of their territorial waters have valid convention certificates. Proof of necessary insurance for oil pollution risks, by a P&I club within the International Group of P&I Clubs, is sufficient evidence to satisfy the certification requirements of the convention.

At this juncture it is important to mention the role P&I associations play in the area of ship owners' third-party liabilities in view of disasters that follow major oil spillage, such as the *Exxon Valdez* in Alaska. A type of insurance, P&I covers most eventualities that are not insured under the normal ship owner's hull insurance. The insurers themselves are P&I mutual clubs, which are associations of ship owners and some charters, who join together to form a club to pay their own and each other's P&I related claims. The advantage of the mutual club system is that it is not run to make a profit, but neither should it make a loss.

Oil pollution coverage is obtained from a P&I club by a ship owner against the risk of pollution of the sea by oil. Traditionally, third-party liabilities of the ship owner have been insured by P&I clubs on an unlimited liability basis. This means that whatever liability the ship owner incurs, unlimited cover will be provided by the respective club. However, given the fact that ship owners are facing ever-increasing oil pollution liabilities involving astronomical clean-up costs, clubs have imposed a limited liability on their coverage for oil pollution, presently standing at five million dollars per incident.

Coverage is given in respect of oil pollution liabilities for losses, damage, costs, and expenses incurred by reason of discharge of oil or any hazardous substance from the entered vessel. This includes schemes such as TOVALOP and other schemes such as Contracts Regarding an Interim Supplement to Tanker Liability for Oil Pollution (CRISTAL). These schemes will be covered in more detail later in the chapter.

In regard to the CLC, P&I clubs issue a *blue card* to a relevant authority; this certifies that the vessel in question is fully covered in accordance with the convention requirements. The competent authority, on presentation of the blue card will issue a convention certificate, which must then be carried at all times on board the particular vessel.

The CLC covers the damage resulting from the actual escape or discharge of oil from a ship as well as the costs of taking preventive measures to mitigate resultant damage. Such preventive measures would include the provision of oil booms following the leakage of oil from a tanker. What is not covered under the convention is the cost of preventive measures taken in situations where there has been no actual spillage of oil.

Under the CLC, a ship owner is strictly liable for pollution damage, unless the owner can prove that the damage:

1. Actually resulted from an act of war, hostilities, or an act of God;

2. Was wholly caused by an act or omission, done with intent to cause damage by a third party;

3. Was wholly caused by the negligence of any government or other authority responsible for the maintenance of navigational aids;

4. Was wholly or partly caused by an act or omission with intent to cause damage by the person who suffered the damage or from that person's negligence, in which case a complete or partial defense may be established.

In the absence of actual fault by the ship owner, under the CLC the owner is entitled to limit liability to special Drawing Rights 133. The International Monetary Fund's (IMF) Special Drawing Rights (SDR) approximates an international unit of account in sterling () so that they could define a currency in terms of SDRs; SDRs' composite exchange rates are not determined by specific trade patterns, but by the general importance of some countries in international trade.

Special Drawing Rights 133, which is approximately $170 per limitation ton, or SDR 14 million (approx $18 million), is the sum at which the ship owner can limit his or her liability, whichever is the smaller sum. In addition, SDRs are treated as fluctuating in value, as they are equal to such a sum, in sterling, that the IMF has fixed as an equivalent of 1 SDR for the day on which the determination is made.

International Oil Pollution Compensation Fund (The Fund Convention)

The Fund Convention is a supplementary convention to the CLC, because only those states that have become parties to the CL.C can become members of the IOPC Fund. The Fund Convention's main functions are to provide supplementary compensation to those who cannot obtain full and adequate compensation for oil pollution damage under the CLC and to indemnify the owner for a portion of his or her liability under that convention.

The compensation payable by the IOPC Fund in respect to an incident is limited to an aggregate amount of sixty million SDR, including the sum actually paid by the owner (or the owner's insurer) under the CLC. Of the forty-six incidents with which the IOPC Fund has dealt, only one has given rise to claims in excess of the limit of compensation that applied to the incident. In all the other incidents, the total amounts of claims arising out of such incidents have been well below the amount of compensation available.

When it comes to ships registered in or flying the flag of a state party to the Fund Convention, the IOPC Fund shall indemnify the ship owner for a part of the aggregate amount of his liability under the CLC. The limit of indemnification payable by the IOPC Fund to the ship owner is 33 SDR for each ton of the ship's tonnage, or 5,667,000 SDR, whichever is less.

In regard to contributions, payments of compensation and indemnification as well as the administrative expenses of the IOPC Fund are financed by contributions levied on any person who has received crude oil and heavy fuel (contributing oil) in a quantity exceeding 150,000 tonnes in one calendar year in a contracting state of the Fund Convention.

Contributing oil is counted for contribution purposes each time it is received at ports or terminal installations in a fund member state after carriage by sea. The term *received* refers to receipt into tankage or storage immediately after carriage by sea. Oil may be imported from abroad, carried from another port in the same state, transported by ship from an offshore production rig or received for transshipment to another port, or received for another further transport by pipeline. The place of loading does not matter.

Now that I have discussed how the fund is financed through contributions levied on any person receiving oil, I can now move to claims. In order for a claim to be accepted by the IOPC Fund, the claim has to be based on a real expense actually incurred and an existing link between the expense and the incident, and an expense that was made for reasonable purposes.

The IOPC Fund pays compensation for expenses incurred for clean-up operations at sea or on the beach (e.g., costs for absorbents and dispersants). It also compensates costs incurred to restore contaminated private property as well as costs for the repair of roads and embankments that have been damaged during the clean-up operations.

For noneconomic environmental damage, the IOPC Fund does not accept claims, but the economic loss suffered by those who depend directly on earnings from coastal or sea-related activities is recoverable from the IOPC Fund, Examples of this include loss of earnings suffered by fishermen and by hoteliers and restaurant owners at seaside resorts.

Protocol 1984

A diplomatic conference, held in London in 1984, adopted two protocols to amend the present CLC and the Fund Convention. The new regimes under the 1984 protocols are referred to as the 1984 CLC and the 984 Fund Convention, respectively. The main amendments adopted by the 1984 conference are as follows.

Higher limits on ship-owner liability were introduced by a special liability limit for small vessels and by a substantial increase in the limitation amounts. The new limitation figures for a ship not exceeding 5,000 gross tonnage was 3 million SDRs; a ship with a tonnage between 5,000 and 140,000 gross tonnage was 3 million SDRs plus 420 SDR for each additional tonnage; and for a ship exceeding 140,000, it was 59.7 million SDRs.

In regard to the 1984 Fund, compensation payable is increased to 135 million SDRs. This includes the compensation payable by the ship owner under the 1984 Civil Liability convention. The limitation figure increases automatically to 200 million SDRs when there are three member states of the 1984 Fund whose combined quantity of contributing oil received during a given year in their respective territories exceeds 600 million tonnes.

It should also be noted that the expenses incurred for preventive measures are recoverable under the 1984 convention, even when there was no spill of oil as a result of the incident, provided that there was a grave and imminent danger of pollution damages. Furthermore, pollution damage of persistent oil, such as slops from unladed tankers, is to be compensated under the 1984 CLC and the 1984 Fund Convention. This is in contrast to the 1969 conventions.

As mentioned earlier, there are also two voluntary industry schemes—TOVALOP and CRISTAL—which provide compensation for oil-pollution damage. These schemes were set up at the same time as the corresponding international conventions were negotiated and are dealt with below.

Tanker Owner's Voluntary Agreement Concerning Liability for Oil Pollution 1969 (TOVALOP)

On October 6, 1969, TOVALOP came into effect when 50 percent of the world's tanker tonnage became parties to the agreement. Today there are some six thousand tankers, combined carriers, and barges entered in TOVALOP, with a total gross tonnage of over 151 million. This figure represents well in excess of 96 percent of the world's tanker tonnage and includes many government-owned fleets. It is rare that any international trading is not entered in TOVALOP.

On a number of occasions since 1969 TOVALOP was amended. In 1978 the scope of the agreement was significantly broadened to bring it into line with the CLC, which came into force in 1975. In 1987 further amendments were introduced that substantially increased the amounts of compensation available to claimants worldwide.

A supplement to TOVALOP contains the new provisions. The agreement that was in force prior to that time is now being referred to as the Standing Agreement. However, the terms of the supplement apply only when a participating tanker involved in an incident is carrying a cargo that is owned by a party to CRISTAL. In all other cases, only the terms and limits of the Standing Agreement remain applicable. Thus, while TOVALOP remains a single agreement, as of February 20, 1987, it has consisted of two tiers; the ownership of the oil cargo is the factor that determines which will apply.

The basis of TOVALOP (which is an agreement entered into by tanker owners and bareboat charters under which the parties agree to assume certain obligations for which they might not otherwise be legally liable) is that when a participating tanker spills persistent oil, the owner or bareboat charter will take appropriate actions. They will also reimburse governments and others who incur reasonable costs in responding to the incident or who suffers pollution damage.

For TOVALOP to apply, it is not a requirement to demonstrate that the tanker owner of bareboat charter was at fault, and there are only a very limited number of circumstances in which a party will be totally free of any obligations under the agreement (e.g., if the incident resulted from an act of war or terrorism). It is therefore possible to obtain compensation without the recourse to legal proceedings that may prove lengthy. It is important to point out here that the TOVALOP party does not waive any rights of recovery from third parties whose fault may have caused or at least contributed to the incidents.

Obligations on a tanker owner or boat charter who is a party to TOVALOP rest on the ability of the party being financially capable of fulfilling its obligations under the agreement. This is normally done by arranging oil-pollution insurance with one of the recognized mutual protection and indemnity associations (P&I clubs) or with the International Tanker Indemnity Association (ITIA) that was set up specifically to provide cover for oil pollution risks.

Contract Regarding a Supplement to Tanker Liability for Oil Pollution (CRISTAL)

The Contract Regarding a Supplement to Tanker Liability for Oil Pollution is the counterpart of the Fund Convention, as opposed to TOVALOP, which is the private industries counterpart to the CLC. The basis of CRISTAL is to provide supplemental compensation in addition to that available from tanker owners and bareboat charters under TOVALOP. It is therefore not surprising to find that many of the definitions and provisions of the CRISTAL contract are complementary to those of TOVALOP, in particular the supplement. However, unlike TOVALOP, claims against CRISTAL are met from an actual fund of money.

For CRISTAL to apply in regard to an incident, a number of conditions must be satisfied. First of all the incident must involve an escape or discharge of persistent oil, or threat thereof, from a seagoing tanker carrying a cargo that is owned or deemed to be owned, at the time of the incident, by a party to CRISTAL.

A shipment of oil will also be considered owned by a party, even though actual title to the oil is in a nonparty, if the shipment is contracted to go to or from a terminal in which a CRISTAL party has ownership or other interest and an incident occurs or pollution damage is caused within 250 nautical miles of that terminal.

Another fundamental condition that must be satisfied before compensation can be obtained from CRISTAL is that the tanker owner or bareboat charter must first pay compensation up to the applicable limit calculated in accordance with the TOVALOP supplement. Like TOVALOP, the CRISTAL contract provides for the reimbursement of reasonable costs incurred by a tanker owner or any other person in taking threat removal measures, preventive measures, or in sustaining pollution damage.

Under the terms of the CRISTAL contract, the following maximum limits of financial responsibility, determined by the gross tonnage of the tanker, can apply. The stated amounts in all cases include compensation that would be payable by the tanker owner, as determined by the limits of financial responsibility under the TOVALOP supplement. For all tankers up to five thousand gross tons, a set maximum of $36 million, for all tankers over five thousand gross tons, $36 million, plus $733 for each gross ton in excess of five thousand gross tons up to a maximum of $135 million.

In the United States the system of liability and compensation for oil spills is rather complex. There are a number of federal statues that provide varying liability standards and scopes of coverage for clean-up costs and damages. For instance, there is the Federal Water Pollution Control Act, the Trans-Alaska Pipeline Authorization Act, the Deepwater Parts Act, and Title III of the Outer Continental Shelf Lands Act Amendments. In addition, thirty-five states have enacted legislation imposing strict liability on the ship owner for clean-up costs.

Furthermore, thirty-two states impose strict liability on spillers for damage to natural resources, and fourteen states impose strict liability for damage to third parties. In many instances, this strict liability is unlimited: thirty-two states for clean up costs, thirteen for third-party damages and twenty-nine for natural resource damages. Three western states—Alaska, Washington, and California—either currently impose liability on the cargo owner or have such legislation pending.

Exxon Valdez

The Exxon Valdez incident in March 1989 gave a new impetus to enact comprehensive oil spill legislation. This led to the Senate and the House of Representatives to pass two bills—S.B. 686 and H.R. 1465. These were considered by a conference committee of the congress that led to a single unified bill that was passed and signed by the president.

In general, the final bill increases the liability of ship owners who trade to the United States, and, in certain circumstances, this liability can be unlimited. On the other hand, the bill does not place any liability on cargo owners, terminals, or cargo users. Furthermore, the bill creates a new federal trust fund for financing clean-up operations, paying oil spill damages, and also mandates new preventives measures, for example, in the fields of ship construction and contingency planning.

A detailed consideration of the bill is not covered in this book. However, it is relevant to examine the position that this places the United States in respective to the 1984 Protocols to the CLC and Fund Conventions. The U.S. delegation at the 1984 conference played a key role in formulating the protocols, particularly as regards the generally agreed upon limits of liability. It has, therefore been a disappointment to the international community that the United States has not, to date, agreed to ratify them and therefore encourage their acceptance by other countries.

One of the main obstacles to U.S. ratification is preemption of States' rights. The House bill H.R. 1465 contained a compromise (the so-called joinder amendment) in order to try to meet this objection, whereby the federal fund would step in to protect the integrity of the treaty regime without the necessity of preempting state law. The Senate bill contained no provision to permit implementation of the protocols and it thus appears that the United States will not do so. It is felt by many that if the United States fails to ratify the protocols, the likelihood of the fund protocol in particular coming into force in the foreseeable future is remote. This encourages other countries to take unilateral action to develop their own oil-spill legislation, leading to a proliferation of differing laws around the world and creating uncertainty in the treatment of oil-pollution claims and liabilities. This is in contrast to the international compensation system that has served claimants as well as ship owners, operators, and cargo owners well for many years.

In the shadow of these historic developments, the Oil Pollution Act 1990 was signed by President George H. W. Bush on August 18, 1990. This imposed new and increased liabilities at the federal level on ship owners and operators for removal costs and damages resulting from spills or threatened spills. Not surprisingly, the world's shipping scene was concerned, and the 1984 Protocols to CLC and the Fund Convention had been rejected.

This resulted in individual states in the United States remaining free to enact their own legislation on oil-pollution liability to complement or supplement federal law. The International group of P&I Clubs expressed great concern about the possibility of increased claims.

In particular, the definition of pollution damage covers a wide range of economic losses and natural resource damages. Substantial penalties can also be imposed that are not in any event subject to limitation. However, limitation rights are perhaps the most important issue and concern, particularly to tanker owners and operators, in light of the Exxon Valdez oil spill and its two billion dollar clean-up costs.

The international system of compensation provided by the two voluntary agreements and two international conventions is unique in the field of environmental pollution and has been found to work to the benefit of the community at large. However, the developments in the United States have now given cause for concern as this opens the possibility of a number of countries developing their own oil spill liability legislation at the expense of the 1984 Protocols.

9

THE SHIP

This chapter takes a closer look at the construction and workings of a modern tanker. It is only appropriate to examine the tanker closely; these magnificent ships are not only the biggest ever built but they also play a fundamental role in international politics and the global energy strategies of the world. Countries have gone to war to ensure that they continue crossing the oceans (e.g., Suez crisis of 1956).

First, I will look at the current design of tankers. The vast majority of oil tankers today have their bridge and machinery located aft, and it is this design we shall concentrate on. Today's tanker is divided into two sections. The forward section in front of the bridge consists of the cargo and ballast tanks while at the after end of the vessel is concentrated most, if not all the machinery.

Generally speaking, the after end of an oil tanker is no different from that of dry cargo vessels. However, certain important items, such as boilers for cargo heating and installations for inert gas are peculiar to tankers. Propulsion power is generated by driving turbines. To produce adequate steam for propulsion of a large tanker such as a VL/ULCC the boiler capacity has got to be tremendous (twin boilers, each with an output of around 50,000 KG per hour/ 110,000 Lb/h on a VLCC is usual) and these boilers must also provide sufficient surplus steam to drive all the machines on board.

Once on the move, however, navigating these giants of the seas requires considerable skill to control a vessel that, due to its size, is affected by forces that arise due to the Earth's rotation. Large objects moving on the Earth's surface are affected by the global spin; this effect is known as the Coriolis force, named in honor of the nineteenth century French mathematician who first described it accurately. The Coriolis force causes a clockwise drift in the northern hemisphere, while south of the equator a moving object is pulled to the left.

Furthermore, steering these large vessels can be a formidable task. For instance, stopping instantly is impossible; it takes at least three miles and twenty-two minutes to stop a 250,000-ton vessel doing sixteen knots. At very low speeds a large tanker may be unable to maneuver at all.

Experiments conducted by the Ship Division of the National Physical Laboratory found that VLCCs and ULCCs are affected by shallows even when they have sufficient water under their keels. Maneuverability is compromised once the water under their keels becomes the equivalent of 40 percent of their draught

So when a tanker of fifty-foot draught still has twenty feet of water under the keel the turning circle is doubled; that is, when its helm is put over the turn, the turn is only half as tight as it would in deeper water. What is particularly worrying is that the tanker's compass continues to indicate that the ship is turning normally when in reality the ship is making a far bigger circle than the master thinks.

When water under a VLCC is down to three feet the ship becomes uncontrollable. This is complicated by the fact that VLCCs tend to squat by the bows, and in ships over two hundred thousand tons this can be as much as three to four feet; this squat is worsened by quite ordinary wave effects, tidal surges, and currents. This therefore increases the possibility of grounding in many shallow inshore areas.

Having seen the challenges associated with the size of the supertanker, I now look at the actual design. The modern tanker's design and cargo tank layout evolved from the T2 tanker of World War II. Up to the war, most deep-sea tankers had accommodation and pump rooms amidships and usually a centerline bulkhead.

As the war progressed it became apparent that there was an urgent need to produce a vessel that could not only segregate several grades of oil but that that could discharge its cargo rapidly in the inevitably congested ports in war-affected areas. This raised the issue of where to locate the pump room. Of course, at the time accommodation amidships was normal in both dry cargo and tanker vessels. It was the advent of the centrifugal pump that eventually resulted in the repositioning of the pump room from amidships to the after section of the ship, just forward of the engine room.

This was necessary because of the difficulties experienced in driving such pumps at a distance from the source of power. It is to be noted that the T2 retained its accommodation amidships; the main pump room was aft and was fitted with centrifugal pumps. The modern tanker is obviously far more sophisticated than the T2; nevertheless, the basic design has not changed much over the years.

The T2 tanker of World War II was not the first vessel to depart from the centerline bulkhead division of cargo tanks, it nevertheless established the now-conventional twin fore and aft bulkhead design, providing three tanks across the vessel. A significant number (over five hundred) were built.

Another important design feature is the double bottom. Unlike dry cargo vessels, virtually all tanker designs were without a double bottom. It is obvious that this feature is unnecessary in a tanker since bunker fuel and water ballast can be carried in selected *main* tanks. However, ever since the grounding of the *Exxon Valdez* the world has become more pollution conscious, and tank allocation has changed to take account of international regulations on Clean Ballast Tanks (CBTs) and Segregated Ballast Tanks (SBTs). At the time of this writing, double bottoms in tankers are to become mandatory. This will be done gradually over a period of time. It is clear that a tanker so fitted will cause less pollution if the bottom is ripped open.

When it comes to modern product tankers, the trend is to load cargo only in the center tanks, retaining the wings for clean ballast. The advantage here is that some protection is available against collision and eventual oil spillage from the side tanks, but it also reduces the number of compartments in which the oil cargo can be loaded.

It is now appropriate to take a look at the actual construction of a tanker's cargo compartment. First, cargo compartments are not large boxes without obstructions; to minimize stress and provide structured strength most tankers have frames inside the tanks. As tanker size increase, the need for hull strengthening increases and it is unusual to find a ship with entirely clear space in its tanks.

It goes without saying that strengtheners must be incorporated somewhere; it is not possible to put stiffeners on the outside; they have to be placed inside the tanks. However, a common trend seems to be the strengthening of wing tanks by incorporating frames within, leaving the center tanks entirely free of obstructions, at least on the tank walls.

Each tank on a modern tanker will be equipped with its own tank hatch, which not only provides access for the crew but is also used for ullage and temperature taking. The tanks will also have individual dropline for loading cargo, and suction lines for discharging. The tanks will also have their own venting system because, as the oil cargo is loaded, there must be means for the air or gas already in the tank to escape. These venting pipes must be sufficiently high to clear the gas well away from personnel and must be fitted with one-way valves to avoid any intake to the tanks.

Modern tankers have all their accommodation aft, and most of the cargo piping to the tanks is carried on deck. This makes it easier to carry out maintenance work. Later on I will look at not only aspects of maintaining a tanker but also the actual operations that are involved in loading and discharging. Loading is a relatively simple process, as the necessary pumps are ashore, and all the tanker needs are the shore connections (manifolds) and piping to feed each tank. The ship will nevertheless have the means to ensure that the oil can be diverted into selected tanks, both for trimming purposes and for segregation of different grades, if any.

Discharging is a little more complicated, since the ship has to have the means to pump oil ashore and also ensure that every possible part of the cargo is removed from the tanks. In an ideal situation every compartment should have a separate line, but in some vessels groups of tanks are connected together by gate valves and the wing tanks drain into the center tank from which the cargo pumps extract the oil to the manifolds on deck.

In addition to the main cargo line for each tank mentioned earlier, each tank would have a much smaller line, connected to a separate stripping pump, which takes over from the main pumps when the cargo reaches a low level in the tanks. Usually the main lines on VLCCs are of a larger diameter, sometimes as much as thirty-six inches, the stripping lines will probably be around twelve inches in diameter. These smaller lines are also useful in removing any accumulated tank drainings and washings from the tanks after tank cleaning.

Some of the most important pieces of cargo equipment aboard a tanker are the cargo pumps. A VLCC will have at least four main cargo pumps. Since the 1950s tankers have been equipped with centrifugal pumps, which can be operated by a steam turbine, electric, or diesel motor. Reciprocating pumps that are worked by steam are usually used for the stripping lines. Before and up to World War II reciprocating pumps were used as the main cargo pumps.

Centrifugal pumps have their limitations. For instance, the pump is not self-priming; in other words, it cannot start unless oil is present in the pump itself. Nevertheless they are very efficient and relatively easy to drive when handling large volumes of oil.

One cannot talk about cargo pumps without giving some consideration to the values necessary to segregate both tanks and lines. Obviously it is not possible to access the valves directly. This has got to be facilitated remotely, outside the tanks. For valves in the suction lines, the operating rods are usually extended to deck level, where one can open or close the valves by turning a valve wheel.

However, as tanker sizes increased, so did the tanks. A cargo tank of about twenty-four meters (eighty feet) is not unusual, making the rod system redundant and outdated. Today valves are normally hydraulically controlled from the deck. Before leaving the question of valves, it is essential to point out that the crew open and close the valve at the right time and in the right sequence to avoid both mixing different grades and the possibility of oil spillage.

Manifolds provide the all-important connection with shorelines and are situated on both port and starboard sides amidships. The main manifolds vary in size, but sixteen inches is a normal diameter, and there will probably be four outlets on each side of the vessel. Some older vessels are able to discharge over the stern through a stern line, as certain older berths can only handle a tanker when berthed stern on, that is, at right angles to the quay.

After looking at some of the equipment aboard a tanker it is only appropriate to give a simple example of the activities involved in preparing and transporting crude oil. First, it is highly likely—especially with a VLCC—that a Crude Oil Wash (COW) will have been carried out concurrently with the previous discharge. It goes without saying that some work has got to be carried out in the tanks before a new cargo is loaded.

However, if the tanker is fitted with SBT or CBT, the preparation work will be reduced. Nevertheless, this is dependant on the instructions from the charterers of the vessel. A tanker may be equipped with fixed or portable washing machines (or both) and the selected tanks are washed with clean seawater. Needless to say, all the cargo tanks will have been pumped with inert gas during discharge.

During tank cleaning, the stripping pump should be in operation to ensure that the washings are constantly collected and transferred to the settling or slop tanks. It may also become necessary to gas free some of the tanks for inspection. This is done by the gas ejectors (ventilators), although wind sails may have to be rigged through the tank lids in older tankers.

The main reason for gas freeing some cargo tanks is so that the crew can check and carry out maintenance work on equipment inside the tanks during ballast passages. With cargo tanks cleaned (if required) and the tanker fully ballasted with clean seawater, the ship is ready for the next loading. As a matter of interest, most tankers will have ballast totaling about one-third of their deadweight. In other words, a 280,000-ton VLCC will carry around 93,000 tons of ballast water. It is important to say that this is only an approximation that will more than certainly be exceeded if the vessel encounters heavy seas on a ballast run.

Now that the vessel is ready to load, there is still a lot of work to be done by the crew, particularly by the chief officer, who is responsible for cargo work in most vessels. The chief officer will need to calculate the amount of space the cargo will occupy, given its gravity and temperature. This is usually based on filling each tank to 98 percent capacity to allow for expansion during a voyage.

If the cargo temperature has to be increased by the vessel during passage or, indeed, will be affected by the climate zones through which the ship will pass, then additional space allowance will have to be made. Tables and computer software is available to provide the necessary conversion factors for both gravity and temperature.

Having calculated the space that each grade will occupy, the chief officer must then allocate the correct number of tanks for each, bearing in mind the stresses that the cargo weights will put on the vessel, the safe and satisfactory trim that provides the most efficient performance and stability, and the possible need to keep cargoes segregated.

In a nutshell, two main activities need to be considered when loading a tanker. These are the removal of ballast water, which the vessel has been carrying for safe passage in the loading port, and the loading of the cargo itself. For vessels that are fitted with CBT or SBT it is possible to discharge the ballast at the same time loading commences.

The world is now much more aware of the dangers of pollution. Consequently harbor authorities no longer allow ballast to be discharged into the sea unless they are absolutely certain that it comes from segregated ballast tanks that are clean. In cases where the vessel is not fitted with SBT but has COW equipment it is obvious that the cargo tanks will not be adequately cleaned for the ballast to be discharged over side.

Under normal circumstances the actual loading starts slowly via the ship's deck manifolds straight into the cargo tanks by means of drop lines into the piping system without passing through the pump room. When loading one grade it is possible to systematically spread this throughout the ship or, when loading more than one grade, it may be necessary to switch from one grade to another during loading operations to ensure that the vessel's stresses and trim are correctly observed.

With loading complete, there is the actual voyage, which—barring heavy weather—should not be complicated. So far I have assumed that the cargo is unheated, but it may be necessary to heat the cargo during the voyage or in order to facilitate discharge.

Heating of an oil cargo is normally carried out by circulating steam through heating coils that are installed at the bottom of each cargo tank. Heating the cargo is one thing. Maintaining a cargo temperature during a voyage can cause some difficulties, for instance the cargo in the wing tanks will be more difficult to heat, as it is nearer the outside of the hull against the seawater, particularly in the Northern and Southern hemispheres during their respective winter months.

Increasing and maintaining heat in cargo tanks costs money in bunkers, and these extra quantities must be allowed for when calculating bunkering for the voyage. Inadequate heating on arrival at the discharge port, for whatever reason, may lead to difficulties in discharging the cargo. For example, if a tanker is unable to heat a cargo correctly, wax precipitation will settle on the coils. This wax is difficult to remove.

The discharging of cargo is similar to the loading operation, except that with discharging a tanker is more under the control of the ship's crew that that of the loading as they can open and close all the necessary cargo valves and to regulate both the centrifugal and reciprocating pumps.

Most modern tankers these days are fitted with COW. Any vessel fitted with COW must have an inert gas system that has to be operational at the time of the washing. Crude oil washing can take place during cargo discharge. This is often done in two distinct operations. First, with a top wash (which may take place when tanks are down to approximately a third full), the COW jets are directed at considerable pressure on the exposed areas of the tank, thereby washing down the remaining clinging oil into the bottom third of the tank.

It is normal practice for one of the vessel's cargo pumps to supply oil to the COW machines. It goes without saying that it is fundamental to wash a tank with the same type of crude oil that is in the tank itself. As the level in the cargo tanks decreases the second operation, the bottom wash begins with separate machines that are mounted lower in the tank than those that do the top wash.

The loading and discharging operations summarized above concern the carriage of crude oil. Obviously a greater emphasis will be placed on cleanliness and segregation when dealing with sophisticated products. Furthermore, clean products are highly volatile and great care has to be taken with regard to the condition of the vessel and elimination of all possibilities of electric discharge that could cause an explosion.

This chapter attempted to highlight the workings of some of the equipment aboard a tanker and some of the activities involved in loading and discharging these magnificent vessels of the high seas. These vessels are not just floating boxes; they are fascinating pieces of technology.

Take the *Jahre Viking* for example. This Ultra-Large Crude Carrier is the largest ship afloat. From tip to tip it is 485.46 meters (1,503 feet long) and when fully laden it sits 24.6 meters in the water, too deep to pass through the Suez and Panama canals, and cannot enter most of the world's major ports.

The *Jahre Viking* has the capacity to load 4,240,865 barrels of oil. Surprisingly it is crewed by only thirty-five to forty persons. This crew is in charge of an oil cargo worth $122 million and is separated from the sea by just 3.5 centimeters of steel plate.

With global oil demand rising and prices hitting record levels, large tankers such as the *Jahre Viking* will continue to sail the seas. In part due to China's growing need for oil that has to be transported by sea, freight rates on the VLCCs sailing between the Gulf and Japan averaged $35,000 a day from 1995 to 2004, but have jumped to as much as $135,000 a day.

In the past, major oil companies ordered hundreds of VLCCs. However, after the closing of the Suez Canal and the subsequent downturn of the world economy there were too many tankers and too little demand.

However, many oil industry analysts and ship owners have been caught short by the global growth in oil demand, particularly from China. China became a net importer of oil in 1993. Its consumption is set to increase. 15It goes without saying that, as China's impact on the world's economy grows, the United States will closely monitor these developments. The public face of U.S. foreign policy is former National Security Adviser and now Secretary of State Condoleezza Rice. The geopolitical ramifications of oil continue to dominate the world's stage. It is for this reason that I link a prominent member of the Bush administration and an oil tanker.

Condoleezza Rice served for nine years (1991–2001) on the board of directors of Chevron, and at one time had an oil tanker named after her in her honor. As mandated by a 1990 U.S. law, the *Condoleezza Rice* was built with a double hull and was capable of carrying 130,000 tons of cargo.

A few months after the ship was certified, the *Condoleezza Rice* was sold to CalPetro Tankers, a third-party shipping firm, and then chartered back to Chevron. Since the *Exxon Valdez* oil spillage, this kind of arrangement has become common among large oil companies, which are anxious to distance themselves from major spills.

Tankers are fascinating and play an important role in geopolitical terms. It was therefore not surprising that shortly after Rice left Chevron to become George W. Bush's national security adviser, the tanker was renamed *Altair Voyager*. Chevron's spokesperson said that this was simply "to eliminate the necessary attention caused by the vessel's original name."

In conclusion, we all know that oil consumption occurs mainly in the industrialized West, while oil production takes place mainly in the Middle East, South America, the former Soviet Union, and West Africa. A huge volume of this oil is traded internationally. This oil is moved by two methods: oil tankers and pipeline. Approximately two-thirds of the world's oil trade (both crude oils and refined products) moves by tanker. This book has attempted to introduce and highlight some aspects of tanker operations in the movement of oil. At the time of publishing investigations continue into the oil–for–food scandal in Iraq. Tankers must have been involved at some stage; we can only wait and see how this was done and which ship tankers will be named.

10

EPILOGUE

At 2:00 AM on August 2, 1990, a hundred thousand Iraqi troops began their invasion of Kuwait. This invasion was the first post-Cold War geopolitical oil crisis. Saddam Hussein miscalculated, as the United Nations did what the League of Nations had failed to do in the 1930s—impose an embargo to frustrate an aggressor.

The invasion and the embargo removed four million barrels of oil from the world oil market on a scale equal to the 1973–1979 crises. As in previous crises, governments, oil companies, and ship owners were worried. However, as it turned out, disruptions caused by this crisis were minimal for the speed and effectiveness of the international reaction ensured that the impact on shipping movements and security was kept to a minimum.

Nonetheless, the Gulf Crisis of 1990 and 1991 highlighted the fact that oil is an essential element in nationalistic propaganda and rhetoric, a major factor in world economics, and a fundamental consideration in the conduct of war. The Gulf crisis brought energy security back onto the political agenda, pushing government to focus anew on ensuring supplies.

Joseph Stanilaw, managing director of Cambridge Energy Research Associates, observed and stated that "The politics of supply security on the part of consuming countries, combined with the need to obtain foreign finance, technology, and resources on the part of OPEC and non-OPEC countries alike, may lead to production increases in many countries around the world." The increased production in areas outside the Middle East will transform the tanker trade and may increase the number of maritime nations.

The Middle East may continue to be unstable for many years yet; it therefore becomes imperative to secure alternative sources around the world. Africa, for example, contains 7.2 percent of the world's proven reserves of oil, 76.7 billion barrels more than the proven reserves of North America or the former Soviet Union.

Geography dictates that African oil is traditionally transported by sea rather than through pipelines, and therefore the tanker market will reconstitute itself to exploit Africans' oil. It is important to note that in sub-Saharan Africa and particularly the Gulf of Guinea crude oil production exceeded four million barrels a day in 2000, more than Iran, Venezuela, or Mexico. West Africa exported almost twice as much crude oil to the United States in 2001 as it did to Europe (68.1 million tonnes to the United States; 34.9 million tonnes to Europe).

The importance of West African oil to the United States cannot be underestimated. On January 25, 2002, a seminar entitled "African Oil: A Priority for U.S. National Security and African Development" was held in Washington, DC, and attended by the then Assistant Secretary of State for African Affairs, Walter Kansteiner III, as well as the ambassadors of several African countries.

According to projections by the U.S. National Intelligence Council, the proportion of oil imported to the United States from sub-Saharan Africa will reach 25 percent by 2015, exceeding that from the Arabian Gulf. This raises the concern over the vulnerability of the African maritime transport infrastructure, both as a potential target for terrorist activity and, perhaps even more worrisome, as a potential weapon of mass destruction. Environmental considerations that are increasingly becoming politicized will have to be contended with as well.

The execution of environmental activist and writer Ken Saro Wira in Nigeria has shown that Africans are demanding—and paying with their lives—that the environment became one of the basic tenets of democratic and good governance reform on the continent. Increased exploration, production, and transportation of oil by sea from Africa will bring about new challenges and problems that may have international ramifications. In 2004–2005 the world witnessed sharp price spikes in the oil market that rose to the fifty-dollar mark due in part to disruptions in oil production in Nigeria by environmental terrorists. This is a precursor of things to come. Appropriate regimes and mechanisms have got to be worked out that address governance issues including corruption, environmental and physical security of oil, and transport infrastructure.

SELECTED BIBLIOGRAPHY AND RECOMMENDED READING

Bailey, Martin, *Oilgate – The Sanctions Scandal*. London: Hodder & Stoughton, 1979.

Basic Documents, 1997-1980, U.S. State Department, No. 256, Statement by President Carter to Reporters at the White House, September 24, 1980.

Bulloch, John, and Morris, Harvey. *Saddam's War: The Origins of the Kuwait Conflict and the International Response*. London: Faber and Faber, 1990.

Collier, Peter, and Horowitz, David. *The Rockefellers: An American Dynasty*. New York: Holt, Rinehart, and Winston, 1976.

Payton-Smith, D. T. Oil: *A Study of War – Time Policy and Administration*. London: HMSO, 1971.

Esser, Robert. "The Capacity Race: The Future of World Oil Supply". Cambridge Energy Research Associates Report, 1990.

Fayle, C. Ernest. Seaborne Trade. 4 vols. London: John Murray, 1924

Foley, Paul. "Petroleum Problems of the World War: Study in Practical Logistics". United States Naval Institute Proceedings 50, November 1924.

Greene, William N. *Strategies of the Major Oil Companies*. Ann Arbor, Michigan: UMI Research, 1982

Feis, Herbert. *Petroleum and American Foreign Policy*. Stanford: Food Research Institute, 1944.

Hope, Stanton. Tanker Fleet: *The War Story of the Shell Tankers and the Men Who Manned Them*. London: Anglo Saxon Petroleum, 1948.

L'Espagnol de la Tramerye, Pierre. *The World Struggle for Oil*. Trans. C. Leonard Lesse. London: George Allen & Unwin, 1924

Lewin, Ronald. The American Magic: *Codes, Ciphers and the Defeat of Japan*. New York: Farrer Straus Giroux, 1982.

Mostert, Noel, Supership, New York: Penguin Books Ltd, 1976.

Sreedhar and Kaul, Kapil. *Tanker War: Aspects of Iraq-Iran War*, New Delhi: ABC Publishing House, 1982.

Stanislaw, Joseph. "The New World Oil Order: Strategies for the 1990s". Cambridge Energy Research Associates, May 1991.

Weinberger, Casper. *Fighting for Peace: Seven Critical Years in the Pentagon*. New York: Warner, 1990.

"World Petroleum Production and Shipping: A Post-Mortem on Suez". Rand Corporation, January 2, 1958.

Yergin, Daniel, *The Prize – Epic Quest for Oil, Money & Power*, New York, Simon & Schuster, 1991.

OTHER

American Petroleum Institute Basic Petroleum Data Book
Baltic Exchange
Grampian Television, Oil. 8-part television series, 1981

International Maritime Organization (IMO)
International Parcel Tankers Association (IPTA)
International Tanker Indemnity Association (ITIA)
International Association of Independent Tanker Owners
(INTERTANKO)
McGraw - Hill. *Platt's Oil Price Handbook and Oilmanac*
Middle East Economic Survey (MEES)
Oil Companies International Marine Forum (OCIMF)
Organization for Economic Co-operation and Development
Organization of Petroleum Exporting Countries (OPEC)
Science Council of Canada
Tanker Advisory Service (TAS)
United Nations Conference on Trade and Development
(UNCTAD)
United Nations Security Council Reports
United States Department of Energy
Woodside Energy
World Bank
World Maritime University

INTERNATIONAL TRADE PRESS

Fairplay Daily News
Lyods List
Petroleum Intelligence Weekly (PIW)
Seatrade
Tanker & Bulk Carrier
Trade Winds

Index

www.ingramcontent.com/pod-product-compliance
Lightning Source LLC
Chambersburg PA
CBHW051445280526
45785CB00003B/1436